THE METRIC SYSTEM

A Programmed Approach

Paul F. Ploutz

Ohio University

CHARLES E. MERRILL PUBLISHING COMPANY
A Bell & Howell Company
Columbus, Ohio

Published by
Charles E. Merrill Publishing Co.
A Bell & Howell Company
Columbus, Ohio 43216

ISBN: 0-675-09057-1

Library of Congress Catalog Card Number: 72-85628

7 8 – 79 78 77 76 75 74

Acknowledgements

James A. Proper, Ann Pirog Proper, high school mathematics teachers, Pine Bush, N.Y.; Robert G. Underhill, associate professor of mathematics, University of Houston; Louis F. Sokol of the Metric Association, Waukegan, Illinois.

Particular appreciation is expressed to Lester C. Mills, professor of science education, Ohio University, for directing much of the field testing of *The Metric System* and for helpful suggestions on the manuscript and in the preparation of the appendices.

Printed in the United States of America

PREFACE

"Think metrically." Most of the countries of the world already use the metric system of measurement. One of the reasons that our products are less competitive than other countries is our continued use of the English system of measurement. Whether we are selling wheat and other agricultural products, machinery, chemicals, automobiles, or technical apparatus, our measurement specifications are now "foreign" to most of the other countries of the world. This measurement dilemma is affecting the success of our exports, our balance-of-payments situation, and ultimately, the number of jobs available and the take-home pay of some of the nation's labor force.

Progress is being made. Metric measurement is now used in much of the defense and aerospace industries. Hospitals, the pharmaceutical industry, and some research related industries now use metric measurement. The public schools use metric measurement in high school biology, chemistry, and physics. The elementary schools are starting to use metric measurement.

While our country's scientific community has already gone "metric," the great bulk of the population does not use metric measurement; many people have never heard of it.

In the meantime, other nations continue to convert to metric measurement. Over 90 percent of the world's people use the metric system. The accompanying graph shows in particular the rapidity of conversion within the last fifteen years. Of the more than 100 nations which have adopted the metric system none have ever abandoned it.

Domestically, the housewife, the man in the street, and the child would gain greater rewards by conversion to the metric system. The consumer would profit by the simplicity of price comparisons of packaging at the counter. The routine household tasks of cooking, measuring material, cleaning, or administering of medicine to the family would be simplified and therefore safer. Estimates have been made that the time, effort, and expense of teaching elementary arithmetic in the nation's schools would be reduced by 25 percent.

It is time for you to become proficient in the use of metric measurement by using this manual. Practice problems have been included to help you "think metrically." This skill should help you to convert from the English system and improve your laboratory skills, as well as to interpret weights and volumes at your local supermarket.

In using this manual *with a group, meaningful classroom activities with practical applications are considered essential,* particularly for students new to the metric system. This manual contains most, if not all, of the basic essentials of metric measurement required for mathematics and for science laboratory programs. The mathematics concepts developed in this manual are compatible with modern mathematics programs and concepts.

Advance of Metric Usage in the World

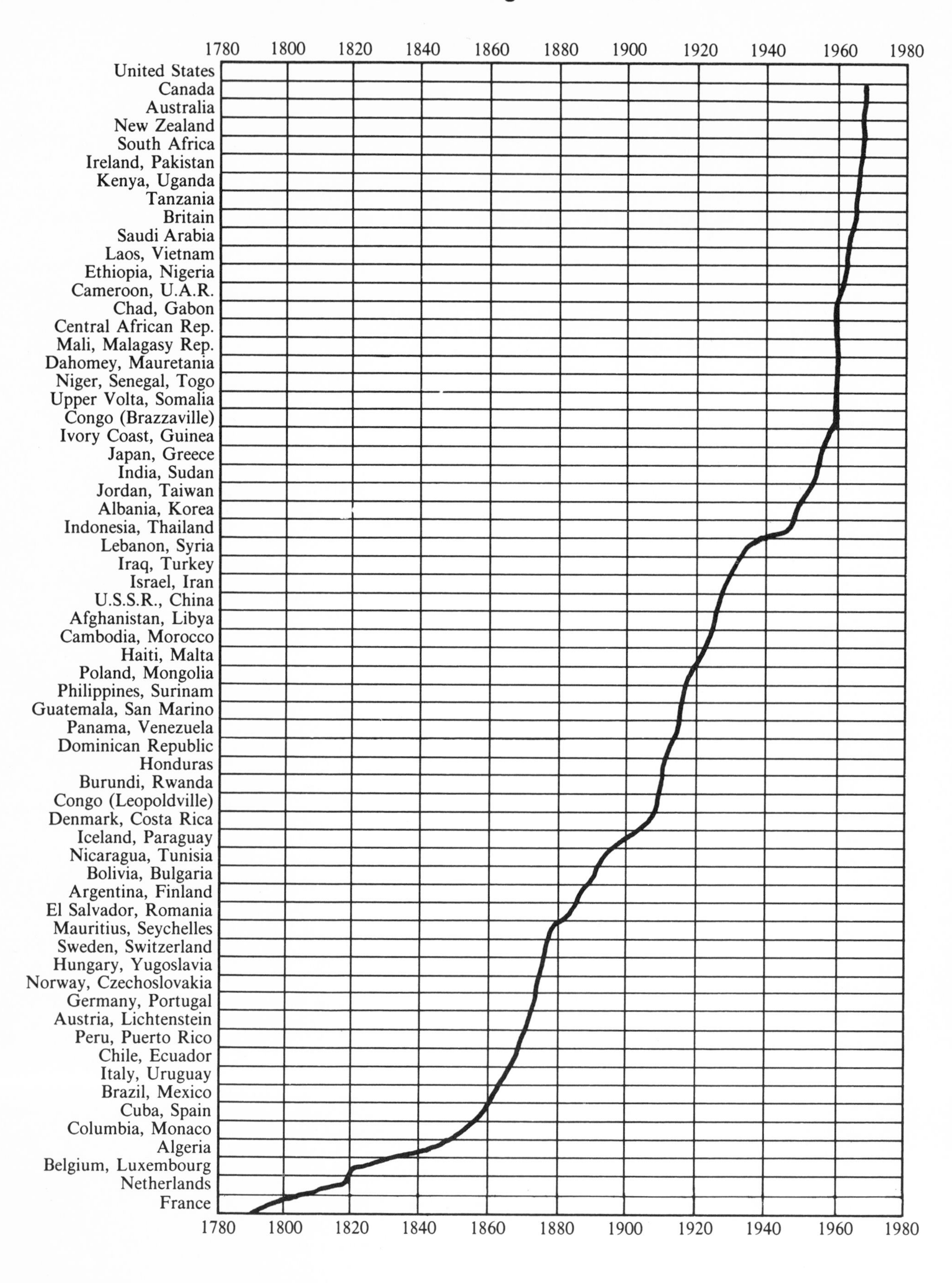

TO THE STUDENT

The following objectives should help you to proceed through each unit of this manual. You should refer to these objectives to keep track of your progress and to make sure you have mastered the concepts of one unit before proceeding to the next.

Unit 1

1. You will be able to solve simple problems using the English system.

Unit 2

1. Through problem-solving experiences, you will find that solving problems in the metric system is easier than solving problems in the English system.
2. You will be able to state how the metric system originated.
3. You will be able to list the prefixes for linear measurement.
4. You will be able to measure objects with a metric rule.

Unit 3

1. You will be able to define the relationship between the linear prefixes milli, centi, deci, deka, and kilo.
2. You will be able to use linear prefixes in problem solving.
3. You will solve simple problems in converting distances from one prefix to another prefix.

Unit 4

1. You will be able to list the symbols for metric units and prefixes: m; km; mm; dm; cm.
2. You will work ratio problems by using metric units in the numerator and denominator.

Unit 5

1. You will be able to select the proper units to measure short and long distances.
2. You will be able to use ratios and prefixes in solving problems.

Unit 6

1. You will be able to convert linear units and prefixes from one unit to another.
2. You will solve problems using the techniques of establishing ratios, simplifying numerators and denominators, and conversion.

Unit 7

1. You will be able to define metric terms and symbols for weights and volume.
2. You will be able to define the term mass.
3. You will be able to define correct units and prefixes for mass; milligram (mg); gram (g); kilogram (kg).
4. You will be able to list correct unit and prefixes for volume; cubic centimeter (cm^3); milliliter (ml); liter (l).

Unit 8

1. You will be able to solve the problems in Unit 8 using a variety of skills you developed in Units 1–7.
2. You will be able to complete all or most of the problems in the Proficiency Text, Appendix C, page 103.

CONTENTS

THE ELEVATOR TO SUCCESS ISN'T RUNNING...

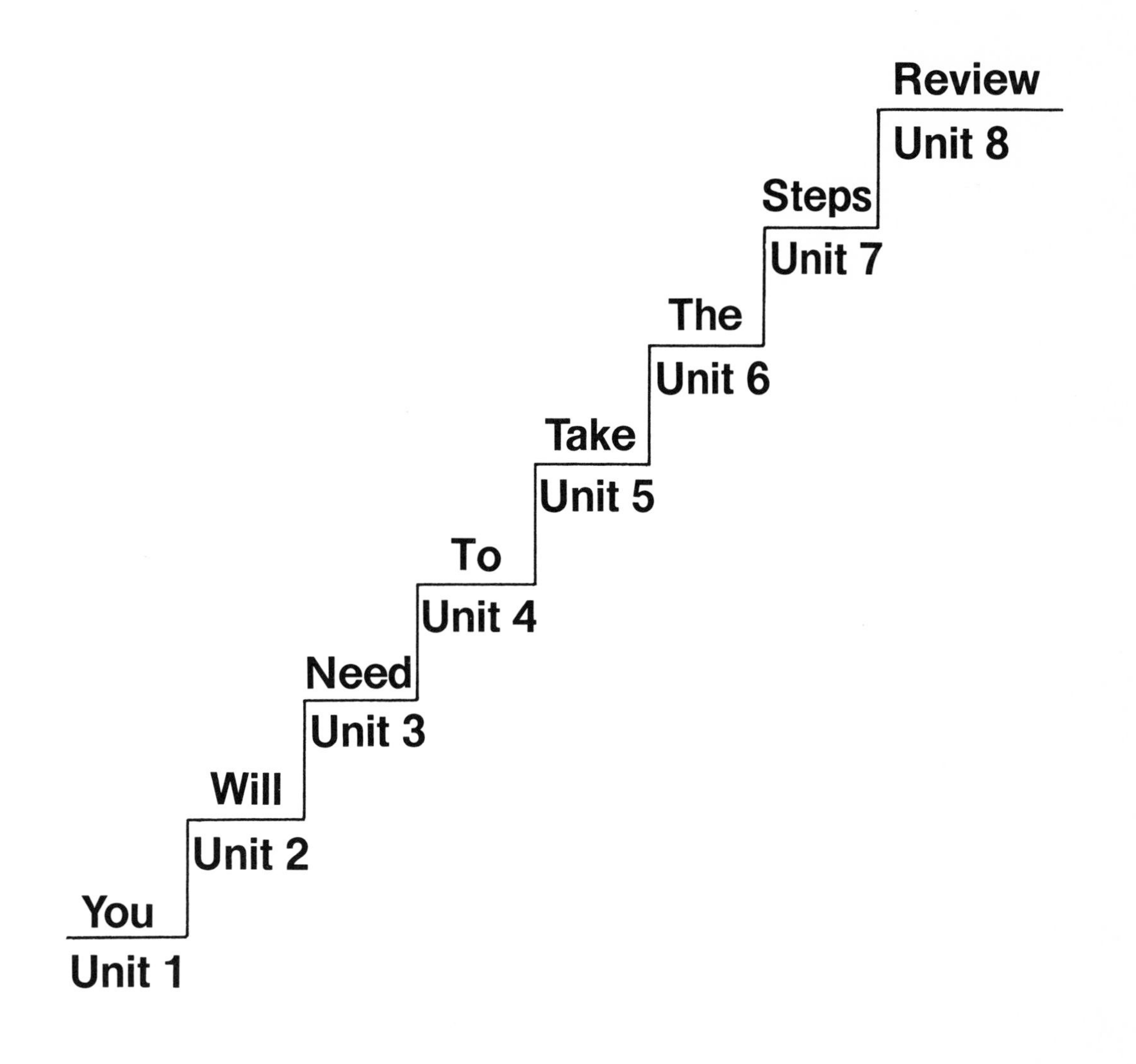

Each unit of *The Metric System* is designed to build your skill in using metric measurement. Each unit builds upon understandings developed in the previous unit and is a step toward competency. In addition to developing competency with the metric system, students pursuing further study will realize they have developed the fundamentals for dimensional analysis.

THINK METRICALLY

This Box is 4 x 15 Centimeters

Directions

This metric manual will help you learn the skills necessary to understand the metric system of measurement. By completing this manual you will rapidly learn the fundamentals of measuring with metric units. You will need to follow these directions carefully:

1. Prepare an answer sheet on regular lined paper. You will number your answers as you complete each question, which is called a *frame.* Many questions have more than one answer. Keep your answer sheet for each unit to record your progress.

2. Read frame 1. You will quickly realize the answer is "English." Write "English" on your sheet as the answer to frame 1.

3. Turn to the end of the unit to find the correct answers to each frame in the unit. You may wish to keep your finger at the page to locate the correct answer easily. Odd numbered questions are on one side, even answers on the other side of the page. Check your answer each time after writing it on your answer sheet. If you have the correct answer, continue to the next frame.

4. If you make an incorrect answer on your sheet, do not erase it. Write the correct answer after it and circle the correct answer for emphasis.

5. If you do not understand why an answer is correct after rereading the frame a second time, ask your teacher for assistance.

6. In Unit 2 you will start using a small metric rule (frame 56). Locate one now. A meter stick will be too long, so you should find a shorter rule.

Students who have used this book suggest a short pause between working with each unit. It is intended that the contents of this manual *be completed in several intervals rather than in a single session.*

Laboratory practice exercises are included in Appendix A, page 97, and are also recommended for practical experience between units. "Think metrically" as often as you can. How tall are you, how much do you weigh, etc.? Practical daily applications of the material you use in this manual will add interest, help you retain what you learn, and help you become effective with metric measurement.

At the end of this manual you will have an opportunity to determine your overall competency in using metric measurement. A proficiency test, Appendix C, page 103, with directions, is provided for your use.

How Imperial Measures Began

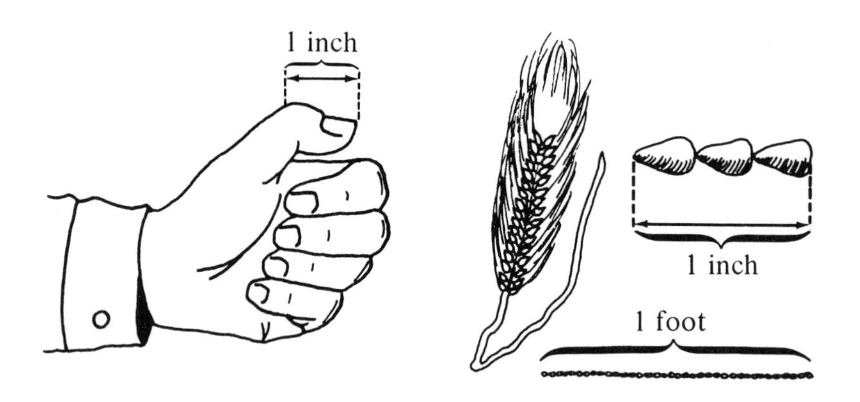

The inch was the knuckle of the thumb, and the foot was 36 barleycorns laid end to end.

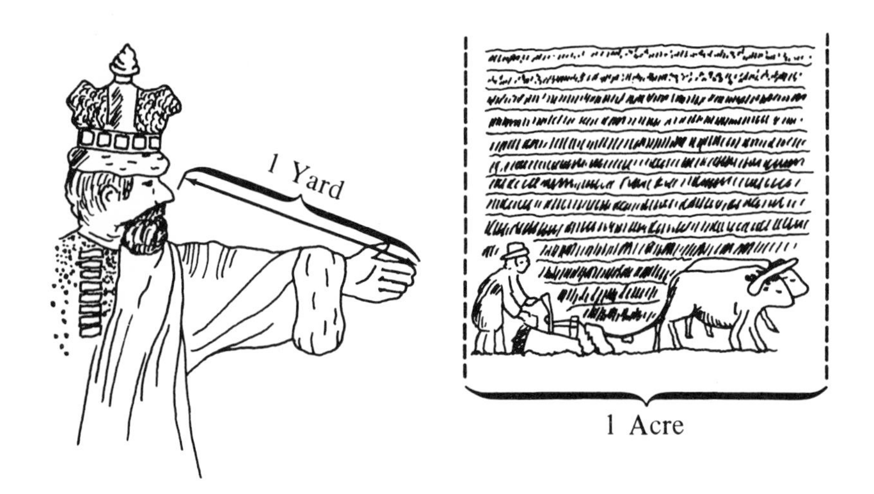

The yard was the distance from the tip of King Edgar's nose to the end of his middle finger, and the acre was the amount of land plowed by a yoke of oxen in a day.

Unit 1

THE ENGLISH SYSTEM

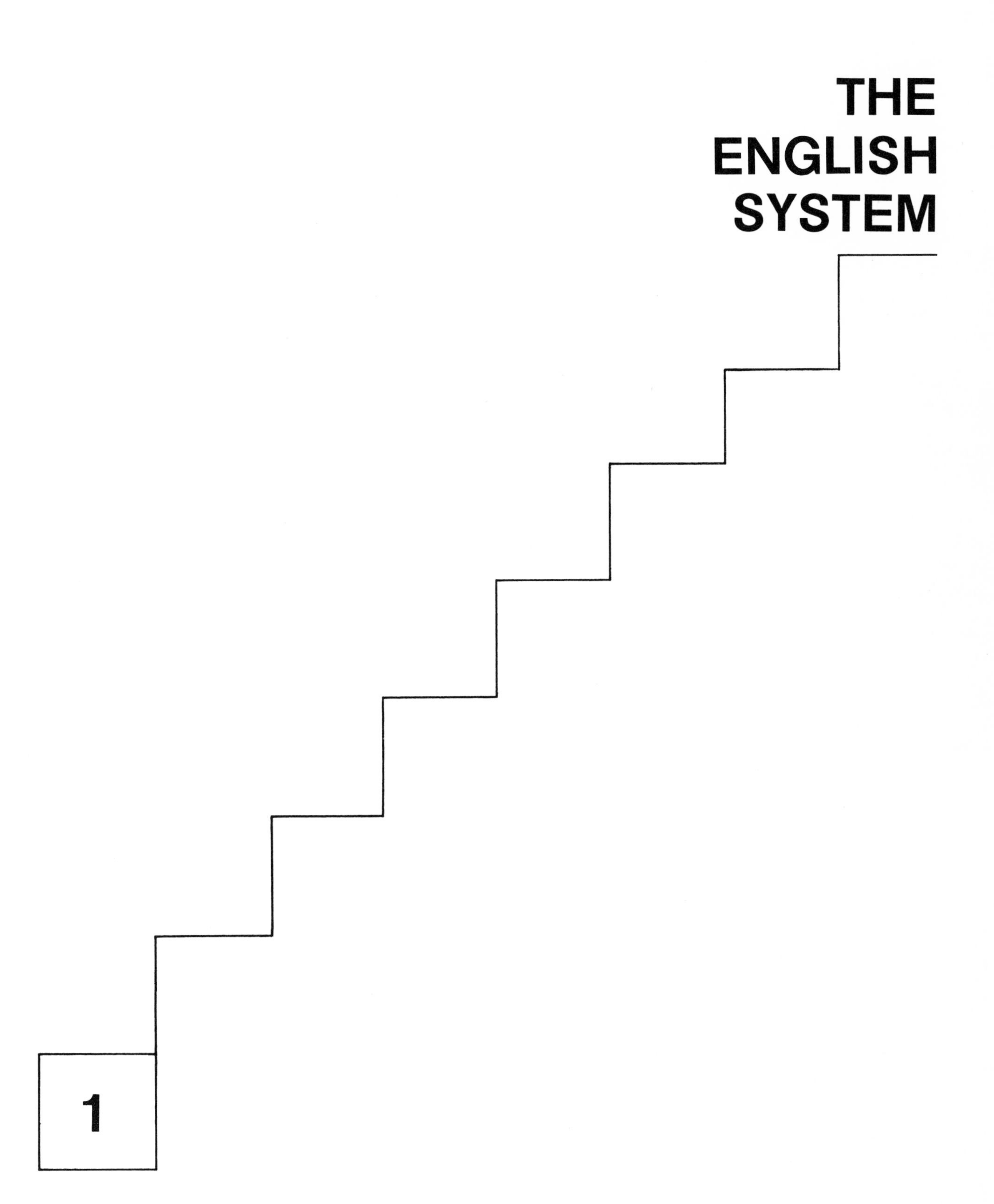

Introduction

Conversion Practice With Proverbs

Old Answers	New Proverbs
1. ____________ ____________	"Give them a *centimeter* and they'll take a *kilometer*."
2. ____________	"A miss is as good as a *kilometer.*"
3. ____________	"I wouldn't touch a skunk with a 3 *meter* pole."
4. ____________	"He's all wool and a *meter* wide."
5. ____________	"There was a crooked man and he walked a crooked *kilometer*."
6. ____________ ____________	"A *gram* of prevention is worth a *kilogram* of cure."
7. ____________	"The Texan pulled a rabbit out of a 42 *liter* hat."
8. ____________	"Don't put your light under a 35 *liter* basket."
9. ____________	"Peter Piper picked 10½ *liters* of pickled peppers."
10. ____________	"Oh Thumbelina, what's the difference if you're very small? When your heart is full of love you're 2.7 *meters* tall."

Turn to page 12 for the answers to Unit 1, Frames 1-31.

1. The English system is the commonly used system of measurement in the United States. Examples of familiar units of the E___________ system are: inch, foot, yard, ounce, pound, ton, pint, quart, and gallon.

2. The inch, foot, and yard are units of measurement in the ___________ system. A foot unit is the same length as 12 inches. A yard unit is the same length as ___________ feet.

3. One foot unit is 12 times larger than the inch unit. The relationship of the foot unit to the inch unit is 1 foot to ___________ inches.

4. One yard is 3 times larger than the foot unit. The relationship of the yard unit to the foot unit is 1 yard to ___________ feet.

5. One mile is 5,280 times larger than a foot unit. The relationship of the mile to the foot is 1 mile to ___________ feet.

6. The rim of a basketball goal should be mounted 10 feet above the gym floor. To change this distance measured in foot units to measurement in inch units, you must know how a foot is related to an inch. The relationship of a foot unit to an inch unit is 1___________ to ___________ inches.

7. To change a measurement given in feet to the same distance in mile units, one must know the relationship of the foot unit to the mile unit. This ___________ is not 12 to 1, nor 3 to 1, but 5,280 feet to 1 mile. The three different relationships of inches to feet, feet to yards, and feet to miles must be known before a person can work successfully with the ___________ system of measurement.

8. If the foot unit was 10 instead of 12 times larger than the inch unit, the English system would be improved. If this were the case, 1 foot would equal 10 inches, 2 feet would equal ___________ inches, and 8 feet would equal ___________ inches. Instead of 12 inches to a foot, problems would be easier if there were ___________ inches to a foot.

9. To further improve the English system, the yard unit could be changed from 3 times to 10 times larger than the foot unit. The mile unit could be set as 10 times larger than the yard unit. If we changed the English system in this way, each larger unit would be ____________ times larger than the next smaller unit. The relationship of a unit to the next larger unit would always be 1 to ____________. The number ____________ is easier to use than 12, 3, 36, or 5,280.

10. The English system unit for measuring weight is the pound. The p____________________ unit is divided into 16 smaller units called ounces. The relationship of pound units to ounce units is therefore 1 pound to ____________ ____________ .

11. The English system unit of the ton is a unit equal to 2000 pounds. The relationship of the ton unit to the pound unit is 1 ton to ____________ ____________ .

12. The English unit for measuring liquids is the quart. The quart is divided into smaller units called pints. If 2 pints are equal to a quart, the relationship of the pint to the quart is 2 ____________ to ____________ ____________ .

13. The relationship of the foot unit to the mile unit is 5,280 feet to 1 mile. This means that there are ____________ ____________ in ____________ ____________.

14. The relationship of one unit to another unit can be called a RATIO. The English system of measurement contains many different ____________s of one unit to another. The relationship of 4 quarts to 1 gallon can be expressed as the ____________ of 4 quarts to 1 gallon.

15. Ratios can be written as fractions to make it possible to work with them mathematically. The ratio of 5 to 2 can be written $\frac{5}{2}$. Likewise, the ____________ of 6 to 3 can be written as ____________ .

16. When changing the relationship in word form to a ____________ in mathematical form follow this rule: The unit given first in the word form is written as the numerator (above the line) and the second mentioned unit is written as the denominator (below the line). The relationship of foot units to yard units is $\frac{\text{3 feet}}{\text{1 yard}}$. The relationship of yard units to foot units is ____________ .

17. The ratio of three to five is written as a fraction as follows: ____________ . The ____________ of five to three is written as a fraction: ____________ .

18. The English unit called the gallon contains 4 quarts. The ratio of the gallon unit to the quart unit is* ____________ .

* Note: Write the name of the unit after each number in the numerator and denominator of the fraction.

19. When describing a given measurement it is always necessary to state the number and the name of the unit. For example, a distance of 10 means nothing unless the name of the ____________ is written after the ____________.

20. Any type of measurement must include two items, a ____________ describing how many and a word giving the type of ____________ .

21. A baby's age may be 15 months, a friend's age 15 ____________ and the age of an ancient statue 15 centuries.

22. The English system unit called the mile contains 5,280 feet. The ratio of the foot unit to the mile unit is ____________ .

23. The ratio of the mile unit to the foot unit is ____________ .

24. The relationship (stated as a ratio) of the yard unit to the inch unit is ____________.

25. Since there are so many different ratios to express the relationship of units in the ____________ system of measurement, this system is difficult to use. For example, change a distance measured as 3.5 mile units to a measurement in yard units. ____________

26. Don't feel bad if you made a mistake in finding the number of yards in 3.5 miles. It is easy to make errors in changing from one unit to another unit when working with the ____________ system of measurement. This is due to the fact that this system has so many ____________ to express the relationships of the different units.

27. Another reason the English system is so ____________ to use is that the name for the units inch, foot, yard, and others give no *clue* to the relationship of the unit to other units. If the foot unit were renamed, the new name could be improved to include a clue. The new name "dozeninch" would be a c____________ showing that a dozeninch unit is equal to ____________ inches.

28. The name of the yard unit could also be improved. The new name should include a ____________ showing how the yard unit is related to the inch unit. The new name, tridozeninch, would give a ____________ that this unit is 3 dozen inches, or 36 times larger than the inch unit.

29. The unit called the foot contains no ____________ that there are 12 inches in the foot unit. The new name for the same unit, the dozeninch, contains a clue relating the unit to the inch unit. The part of the new name (dozeninch) that gives the clue is the prefix __________.

30. The ____________ system is difficult to use because:

(1) There are many different relationships or r____________ to remember and work with, for example, $\frac{1}{12}, \frac{1}{3}, \frac{1}{36}, \frac{1}{5280}$.

(2) The names of the English units have no c____________ in the form of prefixes to show their size in relation to other units; there are no "dozeninch" type names.

31. The English system seems relatively easy to use even though it has many different r____________ to remember. Names of the units do not begin with prefixes that give ____________ to the size. The reason the English system seems easy is that Americans have nearly always used it.

As will be shown in Unit 2 of this learning program the metric system of measurement is planned to overcome the weaknesses of the ____________ system.

END OF UNIT 1

ANSWERS

Unit 1 THE ENGLISH SYSTEM Frames 1-31

1. English

2. English
   three (3)

3. 12

4. three (3)

5. 5,280

6. foot
   twelve (12)

7. relationship
   English

8. twenty (20)
   eighty (80)
   ten (10)

9. ten (10)
   ten (10)
   ten (10)

10. pound
    16 ounces

11. 2000 pounds

12. pints
    1 quart

13. 5,280 feet
    1 mile

14. ratios
    ratio

15. ratio
    $\frac{6}{3}$

16. ratio
    $\frac{\text{1 yard}}{\text{3 feet}}$

17. $\frac{3}{5}$
    ratio
    $\frac{5}{3}$

18. $\frac{\text{1 gallon}}{\text{4 quarts}}$

19. unit
    number

20. number
    unit

21. years

22. $\frac{\text{5,280 feet}}{\text{1 mile}}$

23. $\frac{\text{1 mile}}{\text{5,280 feet}}$

24. $\frac{\text{1 yard}}{\text{36 inches}}$

25. English
    6160 (yards)

26. English
    ratios

27. difficult (hard)
    clue
    twelve (12)

28. clue
    clue

29. clue
    dozen

30. English
    ratios
    clues

31. ratios
    clues
    English

...The Old Way

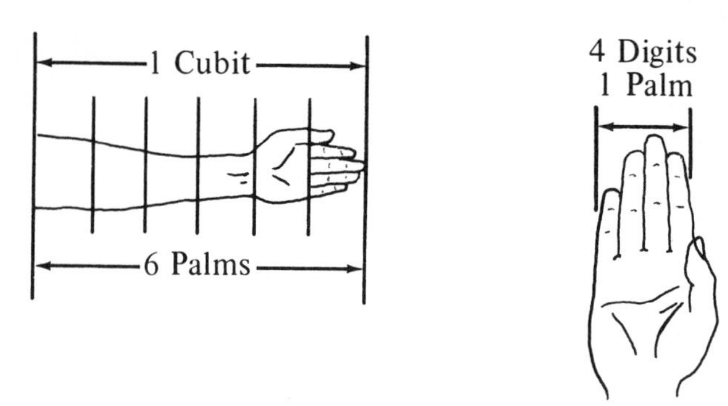

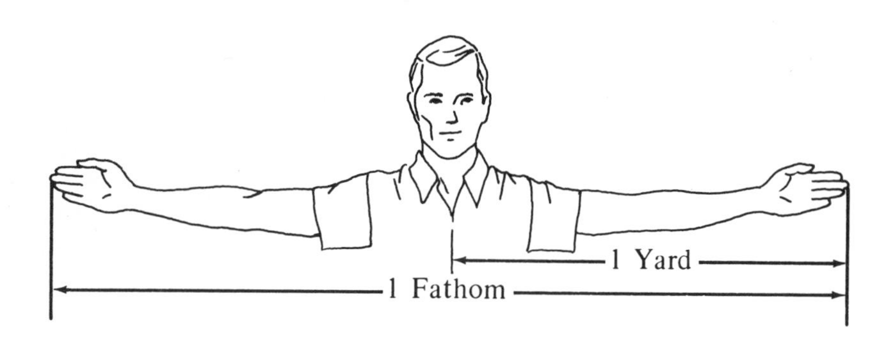

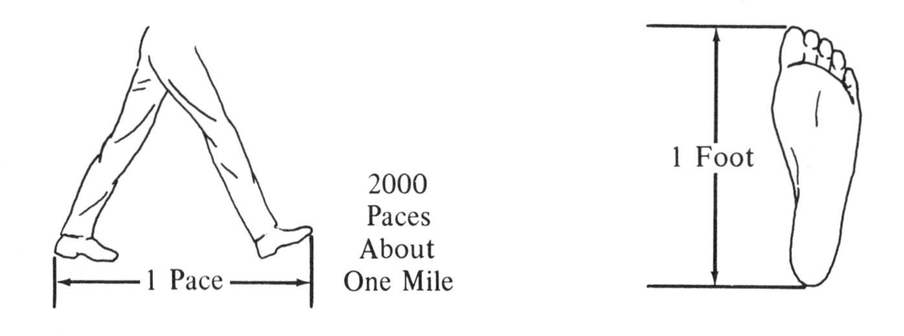

MAN AS A MEASURING ROD

Unit 2

A NEW SYSTEM IS CREATED

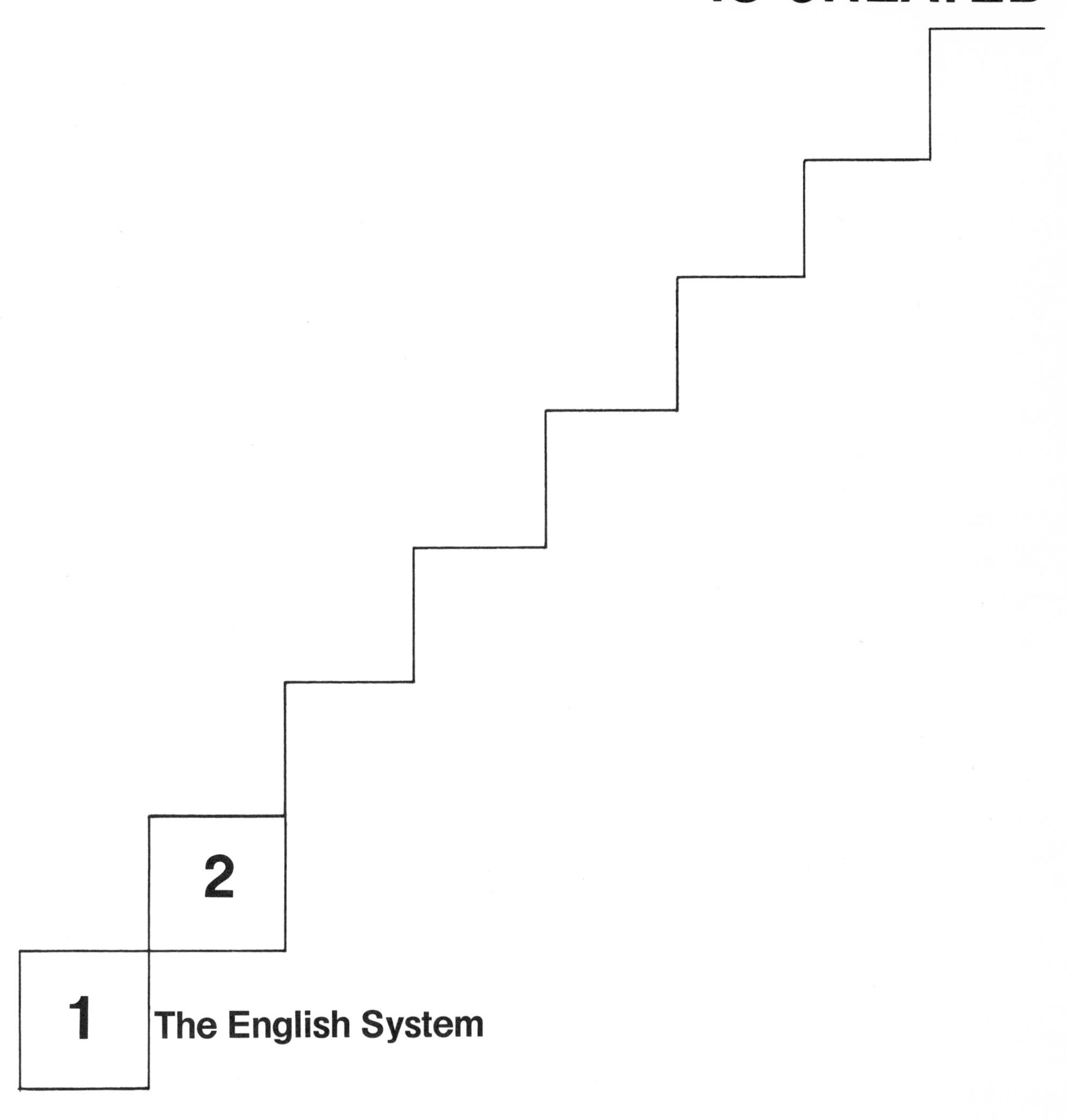

Introduction

Easy Conversions	Difficult Conversions
1 meter	1 yard
or	or
10 decimeters	3 feet
or	or
100 centimeters	4.54 links
or	or
1000 millimeters	36 inches

Advantages of metric system over English system:

1. Name of unit clue to relationship: milli, centi, deci.
2. Convenience; more divisions of meter than of yard.
3. All units are related to each other by powers of 10.
4. Division, addition, subtraction, and multiplication easily done by powers of 10.

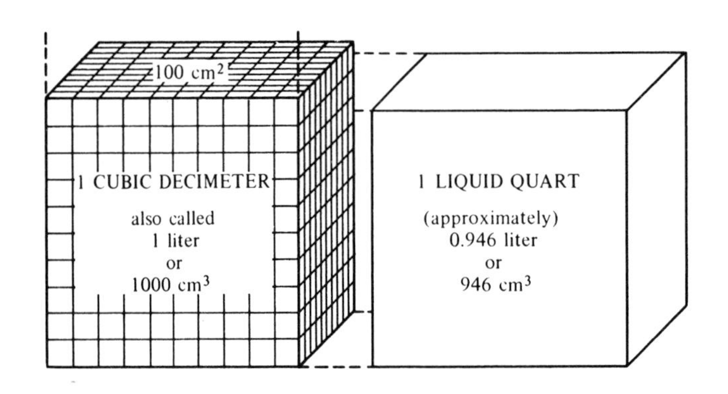

Turn to page 24 for the answers to Unit 2, Frames 32-67.

32. In 1791, during the French Revolution, a committee from the French Academy of Science developed a completely new system of measurement. The *metric* system they produced was carefully planned from the beginning to remove the difficulties of the measuring systems in use at that time. This new system, called the ______________ system, was accepted by the National Assembly of France and made the only legal system of measurement for all of France.

33. Other countries realized the great advantage of the ___________ric system and also adopted it for their offical use. Today, nearly two centuries after the metric system was planned, American medical doctors, scientific groups, many manufacturing companies, the Armed Services, and numerous other groups in the U.S. use the met__________ system of measurement. Despite this fact, the commonly used system of measurement for the U.S.A. is still the difficult to use and unplanned _____________ _____________ of measurement.

34. The standard unit of the metric system upon which other units are based is the *meter*. The name for this unit comes from the French word "to measure." The word is commonly used in this sense when you speak of the water meter, light meter, and the gas___________. To meter something means to _____________ something.

35. The meter unit which was to be the basis for the entirely new measuring system was first calculated to be $\frac{1}{10\ 000\ 000}$ (one ten-millionth) of the distance from the equator to the North Pole. In other words, a line containing _____________ meter units would extend from the _____________ to the North Pole.

Note: The approved metric system no longer uses commas; 1,000,000 would now be 1 000 000. (In the case of four digits the spacing is optional.)

36. After the calculation was made the distance of one meter was carefully marked on a platinum bar to be kept by the French government as a standard for comparison. This official platinum bar, which is about 3 feet long, is still kept by the _____________ government. The platinum bar is no longer the official standard. The international _____________ for the meter is now based on a wavelength of light.

37. Compared to the English system, the length of one meter is almost 40 inches (39.37 inches). For very rough estimates the meter can be considered to be about a yard, or three feet, in length. As a rough estimate, a man who is 6 feet tall will be about ______*______ meters tall. A desk top 20 inches wide is very close to ____________ meter wide.

* to nearest whole meter

38. For very general estimates, a meter can be considered to be only a little longer than ____________ yard, or ____________ feet, and slightly less than ____________ inches.

39. The name of this system of measurement and the name of the basic unit are not spelled the same. The difference is in the word endings "ric" and "er." The met__________ is the basic unit of the met__________ system.

40. You may find the word meter spelled another way in some printed materials (metre). The spelling in this manual is the preferred spelling in this country. Whatever the spelling, the term meter is taken from the French language, and means ____________ ____________.

41. To provide units smaller than the meter, the French Academy scientists divided the meter into 10 equal units called *deci*meters, 100 equal units called *centi*meters, and 1000 equal units called *milli*meters. These three prefixes, "deci" meaning ten, "centi" meaning hundred, and "milli" meaning thousand, were taken from the Latin language. A meter is divided into ten ________meters, one hundred________meters, and one thousand________meters.

42. The first subdivision of the meter into smaller units divides the meter into ____________ equal parts called decimeters. The next subdivision called the centimeter divides the meter into ____________ parts. The millimeter unit divides the meter into ____________ parts.

43. The meter stick found in most classrooms is a convenient and inexpensive model of the standard meter bar kept in France. The meter stick is a meter long and only a few inches longer than the English system ____________ stick. The meter stick is divided into 10 equal sections called ________meters (spelling is like the word *deci*mal).

44. Since the meter stick is approximately 40 inches long (39.37 inches) and since one ______meter is one tenth of a meter, then the decimeter is nearly equal to $\frac{1}{10}$ of 40 inches or ____________ inches.

45.

The line above is one decimeter long. ____________ such units are equal to one meter. The decimeter is divided into 10 smaller units called c______imeters. It takes ____________ of these small units to make a decimeter, and of course 100 small units to make a meter.

46. In a meter there are 10 d____________ each of which contains 10 c____________.

47. A meter contains ____________ decimeters. Each decimeter contains ____________ centimeters. There are ____________ centimeters altogether in a meter.

48. Some people find that a comparison of *d*ecimeters to *d*imes and *cent*imeters to *cent*s assists them in remembering these units. 100 cents make a dollar and 100 ____________ make a meter. Likewise, 10 dimes make a dollar, and 10 ____________ make a meter.

49. Using the money comparison again, 10 *cent*s make a *d*ime and 10 ______meters make a ____________.

50. Four dollars have a value of 400 cents. Four meters contain ____________ centimeters. Eight meters would contain ____________ centimeters and 17 meters would contain 1700 ____________.

51. Every dime is worth 10 pennies. Every decimeter is equal to ____________ ____________.

52. Fifty cents has the same value as ____________ dimes. Eighty centimeters has the same value as ____________ ____________.

53. Leaving the money example, review the size of the meter units covered thus far. One meter is approximately ____________ inches in length and a decimeter is approximately ____________ inches in length.

54. The diameter (distance across) of a nickel is very close to 2 centimeters. The radius (distance from the center to the edge) of a nickel would therefore be equal to 1 ____________.

55. Without digging for your change, draw from memory a circle the size of a nickel. Now draw a line from the center of your circle to the edge to represent the radius. The length of this line measured in metric system units will be ____________ ____________.

56. Now draw a horizontal line 1 decimeter long. You will remember that this unit will be about 4 inches in length. Directly below and parallel to this line, draw ten 1 centimeter lines with their ends touching along a straight line. The ten 1 centimeter lines should extend altogether a distance of one ____________.

57. The relationship of the centimeter unit to the decimeter unit is 10 centimeters to 1 decimeter, or $\frac{\text{? centimeters}}{\text{? decimeter}}$.

58. All units of the metric system are related to each other by some multiple of 10. For example, the relationship between a meter and decimeter is a ratio of $\frac{\text{1 meter}}{\text{10 decimeters}}$. The relationship of the centimeter to the decimeter is ____________ (expressed as a ratio).

59. You may have noticed that the relationship of two units can be expressed as a ratio either of two ways. The relationship of centimeters to meters is ____________, and the relationship of meters to centimeters is ____________. The distance represented by the numerator of the ratio always equals the ____________ represented by the denominator of the ratio.

60. When stating the ratio of one unit to another unit, the d____________ represented by the upper and lower portions of the ratio must be equal. For example, in the ratio $\frac{\text{1 meter}}{\text{10 decimeters}}$ the distance of 1 meter is ____________ to the distance of 10 decimeters.

61. In the ratio describing the relationship of centimeters to decimeters, $\frac{\text{10 centimeters}}{\text{1 decimeter}}$, the distance of 10 centimeters is equal to the distance of 1 decimeter: *but* you will notice that the number by itself above the line *does not* equal the ____________ by itself below the line. It is a fact that 10 ____________ equal 1 ____________ but 10 does not equal 1.

62. 1 meter ____________ 100 centimeters, but 1 by itself ____________* the number 100 by itself. The value of a distance must be expressed in two parts, the number and the name of the unit. The number 15 does not describe a distance because it is not followed by the name of a ____________.

* three words are required

63. It should be clear from this that it is very important *always* to write the name of the ____________ after any number that is used to represent distance.

64. In the last few frames when the relationship of two units of measurement were expressed as a ratio, the actual value of the ratio became equal to 1. The ratio of $\frac{1}{1} = 1$, the ratio of $\frac{4}{4} = 1$, the ratio of $\frac{10}{10}$ = ____________. In each of these ratios the numerator is ____________ to the denominator. The denominator in each case will divide into the numerator exactly ____________ time.

65. The ratio of $\frac{\text{100 centimeters}}{\text{1 meter}}$ is also equal to 1 because the distance in the numerator is ____________ to the distance in the denominator. 1 meter will divide into 100 centimeters exactly ____________ time.

66. $\frac{\text{1 meter}}{\text{10 decimeters}}$ = ____________. $\frac{\text{10 decimeters}}{\text{1 meter}}$ = ____________.

67. The relationship of one type of measuring unit to another type of measuring unit can be expressed in mathematical form as a ____________. The values of the two parts of the ratio are ____________ and the total value of the whole expression is equal to ____________.

END OF UNIT 2

ANSWERS

Unit 2 A NEW SYSTEM IS CREATED Frames 32-67

32. metric

33. metric
metric
English system

34. gas meter
(to) measure

35. 10 000 000 (ten million)
equator

36. French
standard

37. two (2)
one-half (1/2 or .5)

38. 1 yard
3 feet
40 inches

39. meter
metric

40. to measure

41. decimeters
centimeters
millimeters

42. 10
100
1000

43. yard
decimeters

44. decimeter
4 inches

45. Ten (10)
centimeter
ten (10)

46. decimeters
centimeters

47. ten (10)
ten (10)
one hundred (100)

48. centimeters
decimeters

49. centimeters
decimeter

50. 400
800
centimeters

51. 10 centimeters

52. 5 dimes
8 decimeters

53. 40 inches
4 inches

54. centimeter

55. 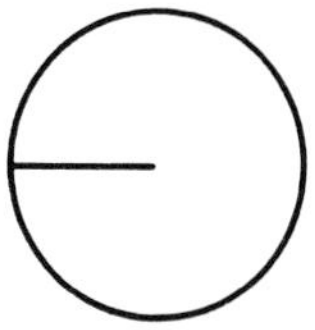 one centimeter

56. decimeter

57. $\frac{10 \text{ centimeters}}{1 \text{ decimeter}}$

58. $\frac{10 \text{ centimeters}}{1 \text{ decimeter}}$

59. $\frac{100 \text{ centimeters}}{1 \text{ meter}}$

    $\frac{1 \text{ meter}}{100 \text{ centimeters}}$

    distance

60. distances
    equal

61. number
    centimeters
    decimeter

62. equals
    does not equal
    unit

63. unit

64. one (1)
    equal
    one (1)

65. equal
    one (1)

66. one (1)
    one (1)

67. ratio (fraction)
    equal
    one (1)

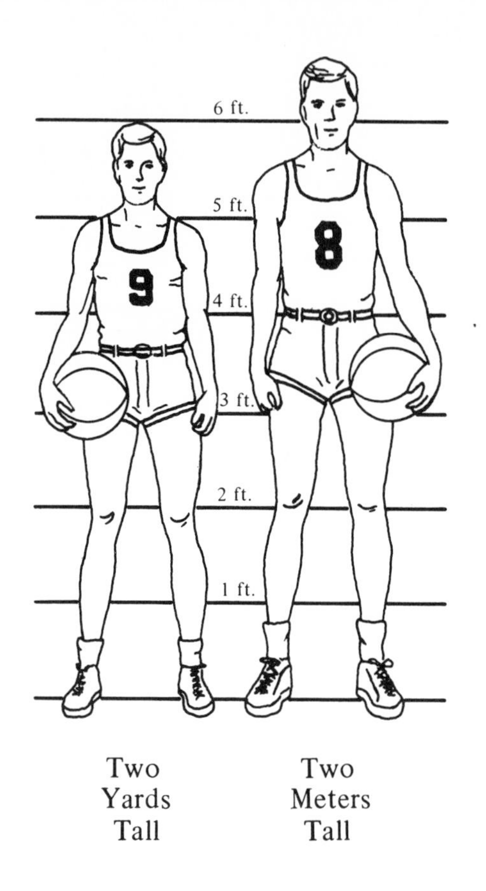

Two Yards Tall

Two Meters Tall

LINEAR CONVERSIONS

1 inch = 2.54 centimeters

1 foot = 0.3 meter

1 yard = 0.9 meter

1 mile = 1.6 kilometers

1 millimeter = 0.04 inch

1 meter = 3.3 feet

1 meter = 1.1 yards

1 kilometer = 0.6 mile

SQUARE CONVERSIONS

1 square inch = 6.5 square centimeters

1 square foot = 0.09 square meter

1 square yard = 0.8 square meter

1 acre = 0.4 hectare

1 square centimeter = 0.16 square inch

1 square meter = 11 square feet

1 square meter = 1.2 square yards

1 hectare = 2.5 acres

Unit 3

ADVANTAGES OF THE METRIC SYSTEM

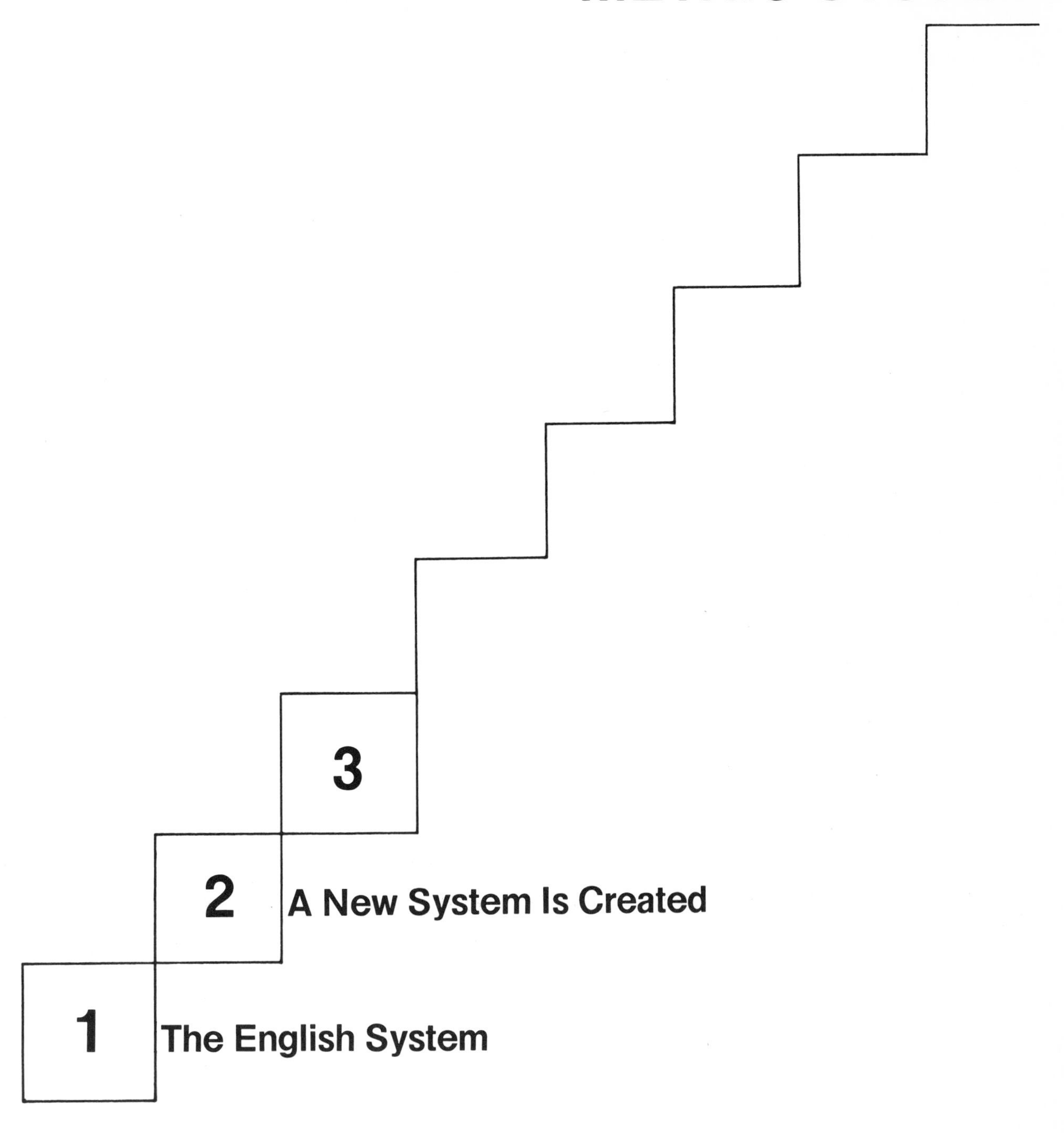

Think Metrically

Box number 2 is one square centimeter (cm^2). Estimate the size of box number 1 and 3. Now check with a rule.

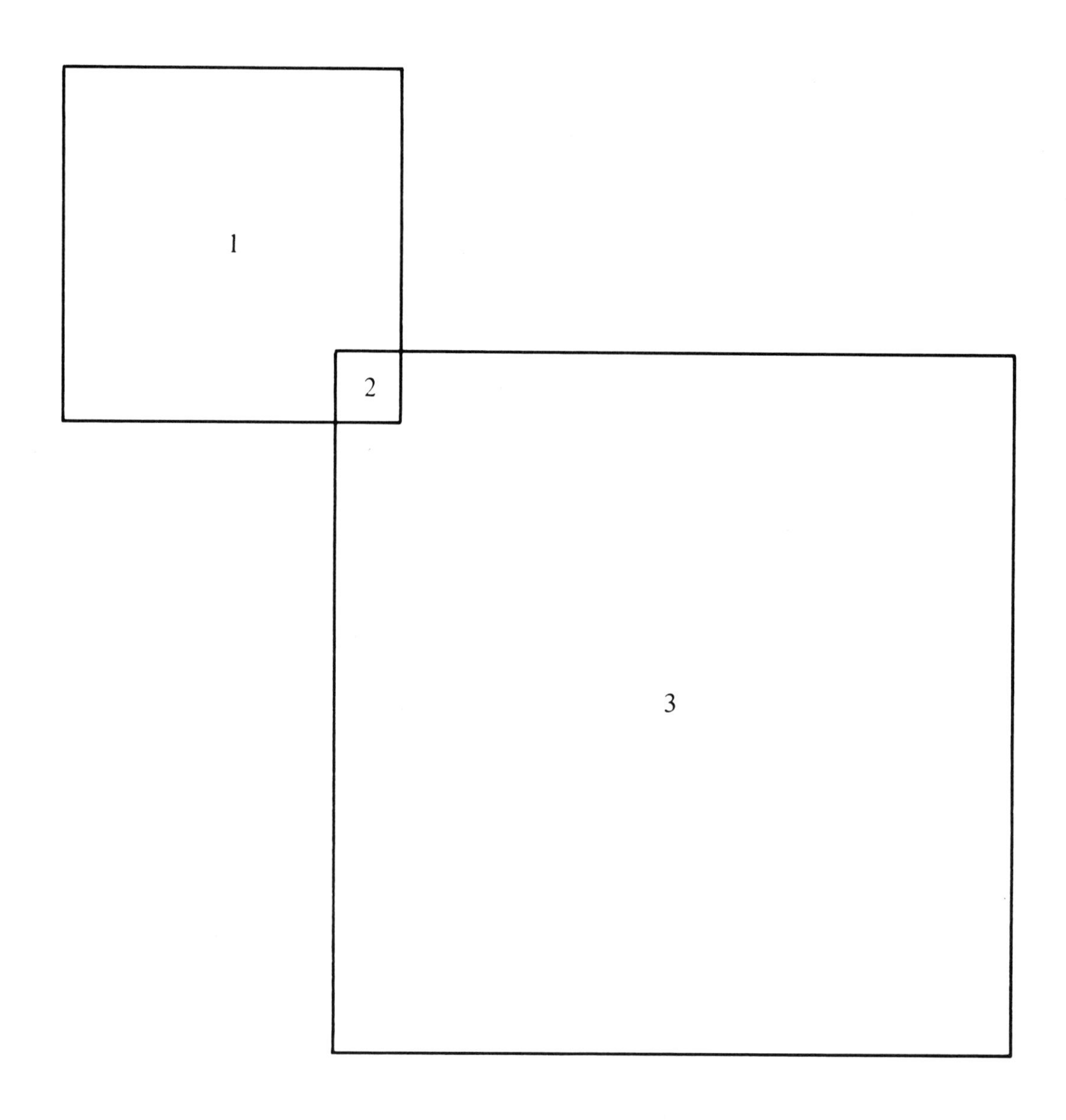

Turn to page 34 for the answers to Unit 3, Frames 68-96.

68. A standard classroom meter stick is subdivided into 10 equal parts to make deci__________ units. The decimeter units are each subdivided into 10 equal parts called c__________meter units. Finally each centimeter unit is subdivided into 10 equal parts called milli__________ units.

69. Ten millimeters is the same distance as 1 centimeter. A distance of 2 centimeters would be equal to 20 __________. A distance of 7 centimeters would be equal to __________ m__________. A distance of 10 centimeters would be equal to __________ __________.

70. If there are 10 millimeters in 1 __________ and 100 millimeters in 10 centimeters, then there are __________ millimeters in 100 centimeters or 1 meter. A meter stick has 1000 very small subdivisions on it called __________.

71. The prefix __________ means $\frac{1}{1000}$. The relationship of the meter to the millimeter expressed as a ratio is __________.

72. In order to estimate the distance of 1 millimeter one can draw a line equal to 1 centimeter which is the radius of a __________. Now divide this line into 10 equal parts. Each of these very small parts would be 1 __________.

73. Another way to estimate the distance of a millimeter is to estimate the thickness of a dime (a dime that has not been worn down on the edges). This might be easier to do than dividing the __________ of a nickel into 10 equal parts. Draw a straight horizontal line 1 centimeter long. On the line mark the thickness of 1 dime, which is also the length of 1 __________.

74. A stack of 20 new dimes would measure approximately __________ millimeters. This would also be equal to 2 __________.

75. Since there are 1000 millimeters in a meter, then a distance of 3 meters would contain __________millimeters. A distance of 5000 millimeters would be a distance of __________ meters.

76. The subdivisions of the meter, the decimeter, the centimeter, and the millimeter are all some multiple of the number ____________. A meter contains ____________ decimeters. A decimeter contains ____________ centimeters. A centimeter contains ____________ millimeters.

77. The ____________fixes milli $\left(\frac{1}{1000}\right)$, centi $\left(\frac{1}{100}\right)$, and deci $\left(\frac{1}{10}\right)$ are all definite clues which give the relationship of the subdivision unit to the basic unit, which is the meter. No such ____________ are found in the ____________ system of measurement.

78. The same well–planned organization of units is present in the metric units larger than the meter. Each larger unit is related to the basic meter unit by a "power" or multiple of ____________. The prefixes for the larger units mean 10, 100, and 1000, but are taken from the *Greek* language instead of the L____________ language.

79. The prefix *deka* (sometimes spelled deca) is found in other words we commonly use such as decade. A decade represents 10 years, and a *deka*meter represents ____________ meters. The preferred spelling in the U.S.A. is ____________meter.

80. The relationship of the dekameter to the meter is 1 ____________ to 10 meters. As a mathematical ratio the relationship of the dekameter to the meter is ____________.

81. The next larger unit is called the hectometer and it represents not 10 but ____________ meters. This unit and the dekameter are used very little in general science. (The English system rod and league units are seldom used today.) The next larger metric unit is frequently used. This unit is the *kilo*meter (only one letter "l"). The ____________ is equal to ____________ meters.

82. The ratio of the meter unit to the kilometer unit is ____________. Four kilometers would be equal to ____________ meters. Fifteen thousand meters would be equal to ____________ kilometers.

83. A kilometer is the metric unit that serves somewhat as the English mile. Automobiles sold in most every country except the U.S.A. have speedometers that are designed to indicate kilometers instead of miles. Since a kilometer is approximately 1/2 mile (actually .62 mile), a car driven 1 kilometer would travel only *about* ____________ mile.

84. Use your rule to draw a horizontal line 100 millimeters long. Draw another horizontal line 62 millimeters long. The 2 lines represent scale drawings of the mile and the kilometer. Label each as follows: the longer line represents the ____________ unit and the shorter line represents the ____________ unit.

85. One mile is the distance of approximately ____________ kilometers. Two miles is about the distance of ____________ kilometers. Ten miles is approximately the same as ____________ kilometers.

86. One kilometer is approximately the distance of ____________ mile. If the distance from one town to another is 20 kilometers, the towns are approximately ____________ miles apart.

87. A person traveling down the highway at 30 miles per hour is moving at approximately the rate of ____________ kilometers per hour. A car speeding along at 100 kilometers per hour is only traveling at approximately ____________ miles per hour.

88. One hundred miles would be approximately ____________ kilometers in length. A 4 kilometer race track would be about ____________ miles in length.

89. You have now learned several important ideas. The relationship of one unit to another can be expressed as a ____________ in the form of a fraction. If the relationship is correctly stated, the value (in distance) of the numerator will ____________ the value (in distance) of the denominator. The actual value of the fraction should always be equal to the number ____________.

90. The relationship of meter units to millimeter units is ____________. The relationship of meter units to kilometer units is ____________.

91. The prefixes used in the metric system for units smaller than the meter are taken from the ____________ language. Prefixes used for units larger than the meter are taken from the ____________ language.

92. A meter is divided into 1000 equal parts called ____________, into 100 equal parts called ____________, and into 10 equal parts called ____________.

93. The name of the unit that represents 1000 meters is the ____________. The length of 1000 meters is approximately ____________mile.

94. A millimeter is approximately the thickness of a ____________. A centimeter is approximately equal to the radius of a ____________. A decimeter is about ____________ inches long. A meter is approximately ____________ inches long.

95. The ____________ system is a *planned* system of measurement making it easier to use than the other system. The advantages are that the names of its units include ____________ to show the relationship to other units. This relationship to other units is always some multiple of the number ____________.

96. Put your ruler away. Now draw a freehand line across your answer sheet from the left to the right side. Mark off and label the distance of:

1 millimeter
1 centimeter
1 decimeter

END OF UNIT 3

ANSWERS

Unit 3 ADVANTAGES OF THE METRIC SYSTEM Frames 68-96

68. decimeter
centimeter
millimeter

69. millimeters
70 millimeters
100 millimeters

70. centimeter
1000
millimeters

71. milli
$\frac{1 \text{ meter}}{1000 \text{ millimeters}}$

72. nickel
millimeter

73. radius |———|
millimeter

74. 20
centimeters

75. 3000
5

76. 10
10
10
10

77. prefixes
clues (prefixes)
English

78. ten (10)
Latin

79. ten (10)
dekameter

80. dekameter
$\frac{1 \text{ dekameter}}{10 \text{ meters}}$

81. 100
kilometer
1000

82. $\frac{1000 \text{ meters}}{1 \text{ kilometer}}$
4000 meters
15 kilometers

83. one-half (1/2)

84. mile
kilometer

85. two (2)
four (4)
twenty (20)

86. one-half (1/2)
10 miles

87. 60
50

88. 200 kilometers
2 miles

89. ratio
equal
one (1)

90. $\frac{1 \text{ meter}}{1000 \text{ millimeters}}$

$\frac{1000 \text{ meters}}{1 \text{ kilometer}}$

91. Latin
Greek

92. millimeters
centimeters
decimeters

93. kilometer
one-half (1/2)

94. dime
nickel
4 inches
40

95. metric
prefixes (clues)
ten (10)

96. Check your marks with a metric ruler.

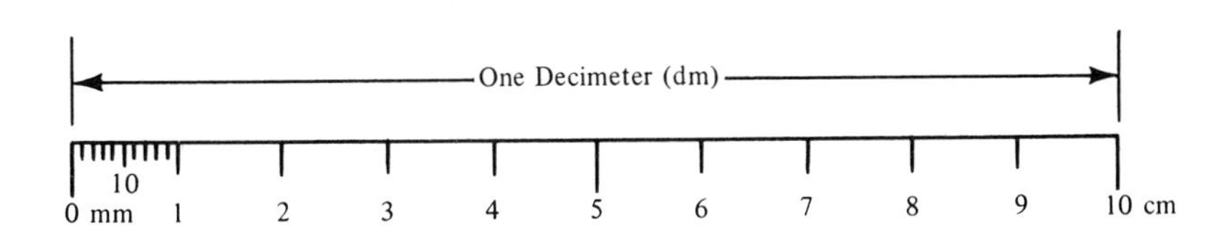

Unit 4

FOUNDATIONS FOR CONVERSIONS

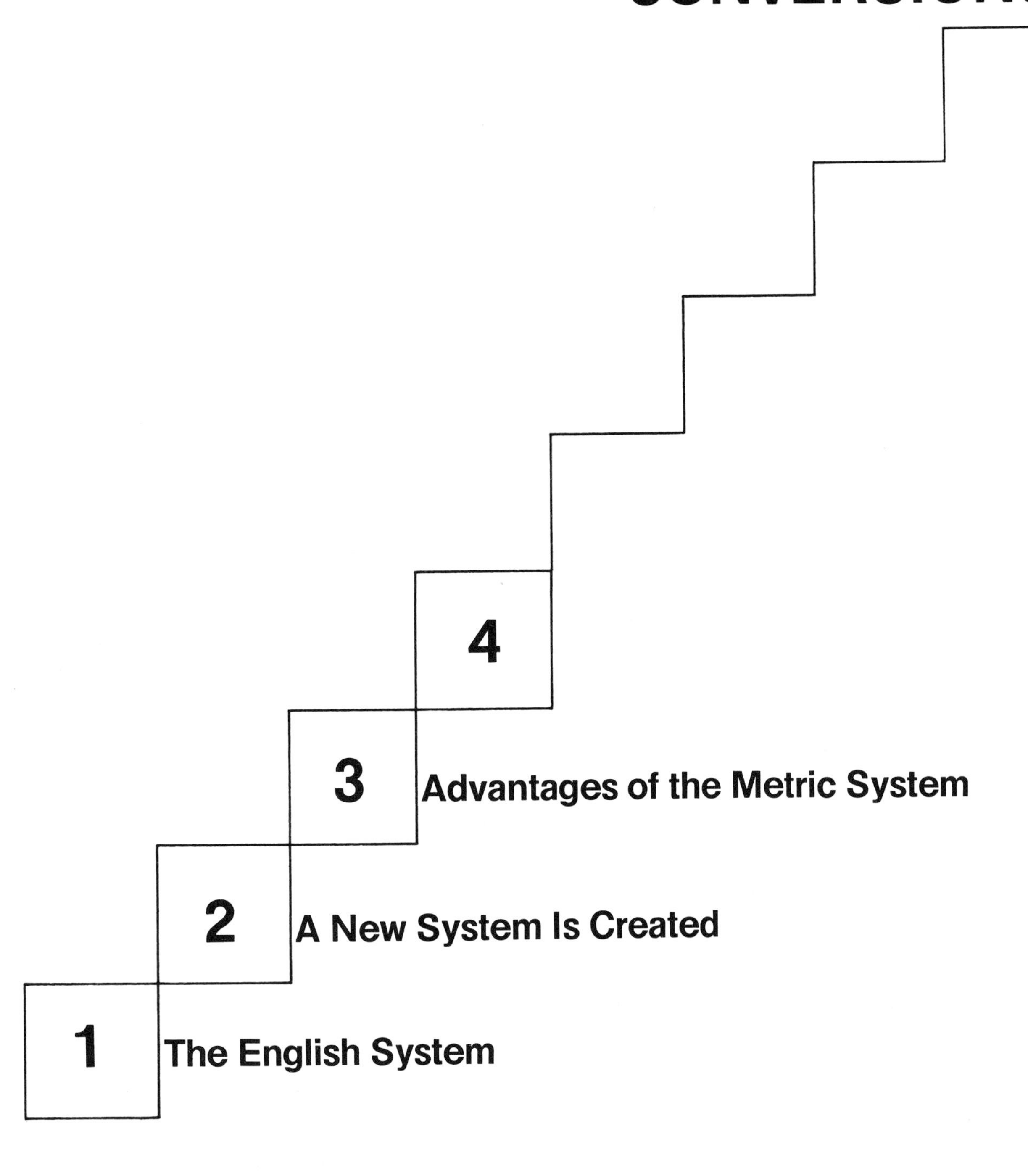

Introduction

Ratio Table

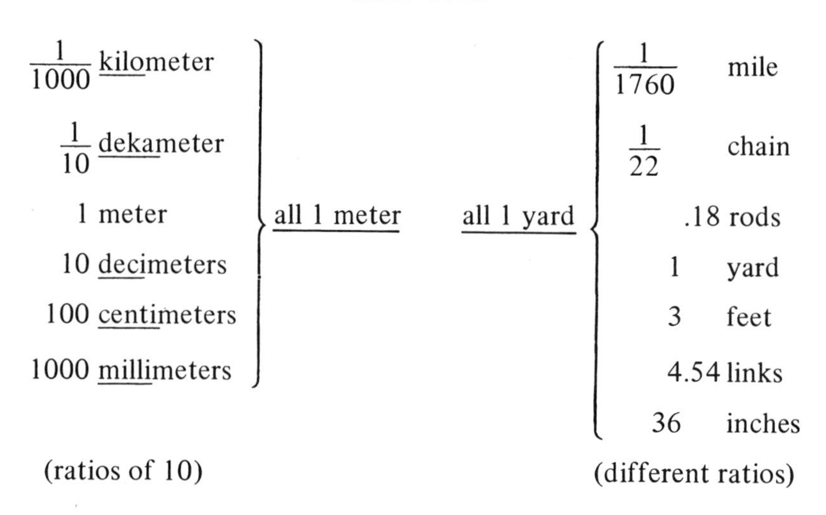

Metric			Customary
$\frac{1}{1000}$ kilometer			$\frac{1}{1760}$ mile
$\frac{1}{10}$ dekameter			$\frac{1}{22}$ chain
1 meter	all 1 meter	all 1 yard	.18 rods
10 decimeters			1 yard
100 centimeters			3 feet
1000 millimeters			4.54 links
			36 inches
(ratios of 10)			(different ratios)

Think Metrically

What is the approximate length and width of this manual in centimeters?

Measure . . . how are you doing?

Turn to page 44 for the answers to Unit 4, Frames 97-133.

97. It will be useful at this point to learn the symbols* for the metric units of distance.
The symbol m stands for ____________.
km stands for ____________.
mm stands for ____________.

* Symbols for units and prefixes are *symbols* and not abbreviations. Many people erroneously refer to them as abbreviations.

98. We will not use dekameter and hectometer. The symbol cm stands for ____________ and dm stands for ____________.

99. In each case the first letter of the prefix (followed by the letter m for meter) forms the symbol.
The symbol for millimeter is ____________.
The symbol for centimeter is ____________.
The symbol for decimeter is ____________.
The symbol for kilometer is ____________.

100. If you are to be successful in working with measurements, you will need to have a clear understanding of some of the rather simple laws of mathematics. The remainder of this lesson will review the concepts you must be able to use.
You have learned that the relationship of meters and kilometers can be expressed as a ____________.

101. The relationship between the two units can be expressed either of two ways: as $\frac{1 \text{ km}}{1000 \text{ m}}$ or as $\frac{1000\ ?}{1\ ?}$.

102. In these ratios the value of the distance represented above the line is equal to the ____________ of the distance represented below the line. When a fraction has the same value in the numerator and denominator, we can say that it is equal to 1. For example: $\frac{1}{1} = 1, \frac{2}{2} = 1, \frac{?}{6} = 1$, and $\frac{9}{?} = 1$.

103. The ratio $\frac{10}{10}$ is equal to 1 because the value above the line equals the ____________ below the line.

104. In mathematics placing one number over another number $\left(\text{like } \frac{4}{2}\right)$ means that the upper number or numerator is to be divided by the lower number or denominator. The symbol ÷ for dividing uses this idea. The two dots represent two numbers and the horizontal line means one number is placed over the other number. $\frac{8}{2}$ means the same as ____________ ÷ ____________.

105. $\frac{10}{10}$ is the same as 10 ÷ 10. Both ways of writing the statement mean one number can be divided into the other number a certain number of times.

$\frac{10}{10}$ or 10 ÷ 10 means the number 10 can be divided into the number ____________ exactly ____________ time.

106. $\frac{10}{2}$ or ____________ ÷ ____________ means 2 can be divided into ____________ exactly 5 ____________.

107. When we say $\frac{5}{5}$ is 1, we mean that ____________ can be divided into ____________ exactly 1 ____________.

108. In a ratio $\frac{1000 \text{ mm}}{1 \text{ m}}$ the value of the distance 1 m can be divided into the value of the distance ____________ mm exactly ____________ ____________.

109. $\frac{1000 \text{ mm}}{1 \text{ m}}$ = ____________. $\frac{1 \text{ m}}{1000 \text{ mm}}$ = ____________.
The answer 1 in each of these cases means 1 ____________.

110. $\frac{1 \text{ km}}{1000\text{m}} = 1$. The actual value of the ratio $\frac{1 \text{ km}}{1000 \text{ m}}$ is 1. This would not be labeled 1 *m* or 1 *km*, but would mean 1000 m can be divided into 1 km exactly 1 ____________.

111. It is not necessary to label a number "times" when the number is the result of division. We know that when such division is carried out the answer will not be labeled but will mean the number of ____________ one value can be divided into another value.

112. When the relationship of one unit to another is expressed as a ratio, the ratio must have an actual value of 1. The value of the distance above the line must ____________ the value of the ____________ below the line in order to give the ratio an actual value of ____________.

113. In the ratio $\frac{1\ \text{km}}{1000\ \text{m}}$ the value of each distance must be written in two parts, the *measure* and the unit of measurement.

$\frac{1}{1000}$ does not equal 1 and $\frac{\text{km}}{\text{m}}$ does not equal 1, but $\frac{1\ \text{km}}{1000\ \text{m}}$ is ____________ to ____________.

114. The ratio $\frac{1\ \text{m}}{10\ \text{dm}}$ is equal to ____________ because the value of the distance above the line is ____________ to the ____________ of the distance below the line.

115. The ratios of $\frac{1000\ \text{m}}{1\ \text{km}}$, $\frac{1\ \text{km}}{1000\ \text{m}}$, and $\frac{1\ \text{dm}}{10\ \text{cm}}$ are all equal to ____________.

116. You will need skill in recognizing ratios. Which of the following ratios is *not* equal to 1: A, B, or C?

(A) $\frac{1\ \text{m}}{100\ \text{cm}}$ (B) $\frac{1}{100}$ (C) $\frac{100\ \text{cm}}{1\ \text{m}}$

117. Which of the following is not correct: A, B, or C?

(A) $\frac{1000\ \text{km}}{1\ \text{m}} = 1$ (B) $\frac{1\ \text{m}}{100\ \text{cm}} = 1$ (C) $\frac{10\ \text{dm}}{1\ \text{m}} = 1$

118. Which of the following is correctly written: A, B, or C?

(A) $\frac{1\ \text{m}}{10\ \text{dm}} = 1\ \text{m}$ (B) $\frac{1\ \text{m}}{10\ \text{dm}} = 1\ \text{dm}$ (C) $\frac{1\ \text{m}}{10\ \text{dm}} = 1$

119. Which of the following correctly expresses the basic relationship of one unit to another: A, B, or C?

(A) $\frac{1\ \text{cm}}{10\ \text{mm}}$ (B) $\frac{1\ \text{cm}}{10\ \text{dm}}$ (C) $\frac{100\ \text{mm}}{1\ \text{cm}}$

120. Which of the following is a correct relationship?

(A) $\frac{\text{5 dimes}}{\text{4 quarters}}$ (B) $\frac{\text{200 cents}}{\text{3 half dollars}}$ (C) $\frac{\text{75 pennies}}{\text{3 quarters}}$

121. A principle of mathematics states that multiplying a numerical value by the number 1 does not change the value.

$6 \times 1 =$ ______. $6 \times \frac{1}{1} =$ ______. 6 meters $\times \frac{1}{1} =$ ______.

122. $4 \times \frac{5}{5} = 4$ $2 \times \frac{10}{10} =$ ______. 2 cm $\times \frac{10}{10} =$ ______.

123. $8 \times \frac{1268}{1268} =$ ______. 1000 m $\times \frac{41\ 248}{41\ 248} =$ ______.

124. $8 \times \frac{10\text{ m}}{10\text{ m}} =$ ______ $\left(\text{Hint: } \frac{10\text{ m}}{10\text{ m}} = 1\right)$. $4 \times \frac{1\text{ m}}{100\text{ cm}} =$ ______.

125. 16 mm $\times \frac{1000\text{ m}}{1\text{ km}} =$ ______. 2 m $\times \frac{1\text{ km}}{1000\text{ m}} =$ ______.

126. Another principle of mathematics states that a numerical value may be changed to a fraction by placing it over 1.

$6 = \frac{6}{1}$. $2 = \frac{?}{1}$. $3 =$ ______. 4 cm $= \frac{?}{1}$.

127. 15 m $= \frac{?}{1}$. Change 100 meters to a fraction: 100 m = ______.

128. $3 \times 1 =$ ______. $3 \times \frac{10}{10} =$ ______. $\frac{3}{1} \times \frac{10}{10} =$ ______.

129. If $\frac{3}{1} \times \frac{10}{10} = 3$, then $\frac{3}{1} \times \frac{1\ \text{m}}{10\ \text{dm}} = 3$ and $\frac{5}{1} \times \frac{1000\ \text{m}}{1\ \text{km}} =$ ____________.

130. Which of these (A or B) has an actual value of 8?

(A) $\frac{8}{1} \times \frac{1\text{m}}{1\ \text{cm}} =$ (B) $\frac{8}{1} \times \frac{1\ \text{m}}{100\ \text{cm}} =$

131. What is the symbol for each of the following?

meter

millimeter

kilometer

centimeter

decimeter

132. What is the relationship of the millimeter unit to the meter unit expressed as a ratio?

133. A measurement of 1000 mm placed over a measurement of 1 m $\left(\frac{1000\ \text{mm}}{1\ \text{m}}\right)$ represents an actual value of ____________.

END OF UNIT 4

ANSWERS

Unit 4 FOUNDATIONS FOR CONVERSIONS Frames 97-133

97. meter
kilometer
millimeter

98. centimeter
decimeter

99. mm; cm; dm; km

100. ratio

101. $\frac{1000\text{ m}}{1\text{ km}}$

102. value; 6; 9

103. value

104. 8 ÷ 2

105. 10; $\underline{1}$ time

106. 10 ÷ 2; 10; times

107. 5; 5; 1 $\underline{\text{time}}$

108. 1000; 1 time

109. 1; 1; 1 $\underline{\text{time}}$

110. time

111. times

112. equal; distance; one (1)

113. equal; 1

114. 1; equal; value

115. 1

116. B

117. A

118. C

119. A

120. C

121. 6; 6; 6 m

122. 2; 2 cm

123. 8; 1000 m (1 km)

124. 8; 4

125. 16 mm; 2 m

126. $\frac{2}{1}$; $\frac{3}{1}$; $\frac{4\text{ cm}}{1}$

127. $\frac{15\text{ m}}{1}$; $\frac{100\text{ m}}{1}$

128. 3; 3; 3

129. 5

130. B

131. m; mm; km; cm; dm

132. $\frac{1000\text{ mm}}{1\text{ m}}$

133. one (1)

Unit 5

MECHANICS
OF
MEASUREMENT

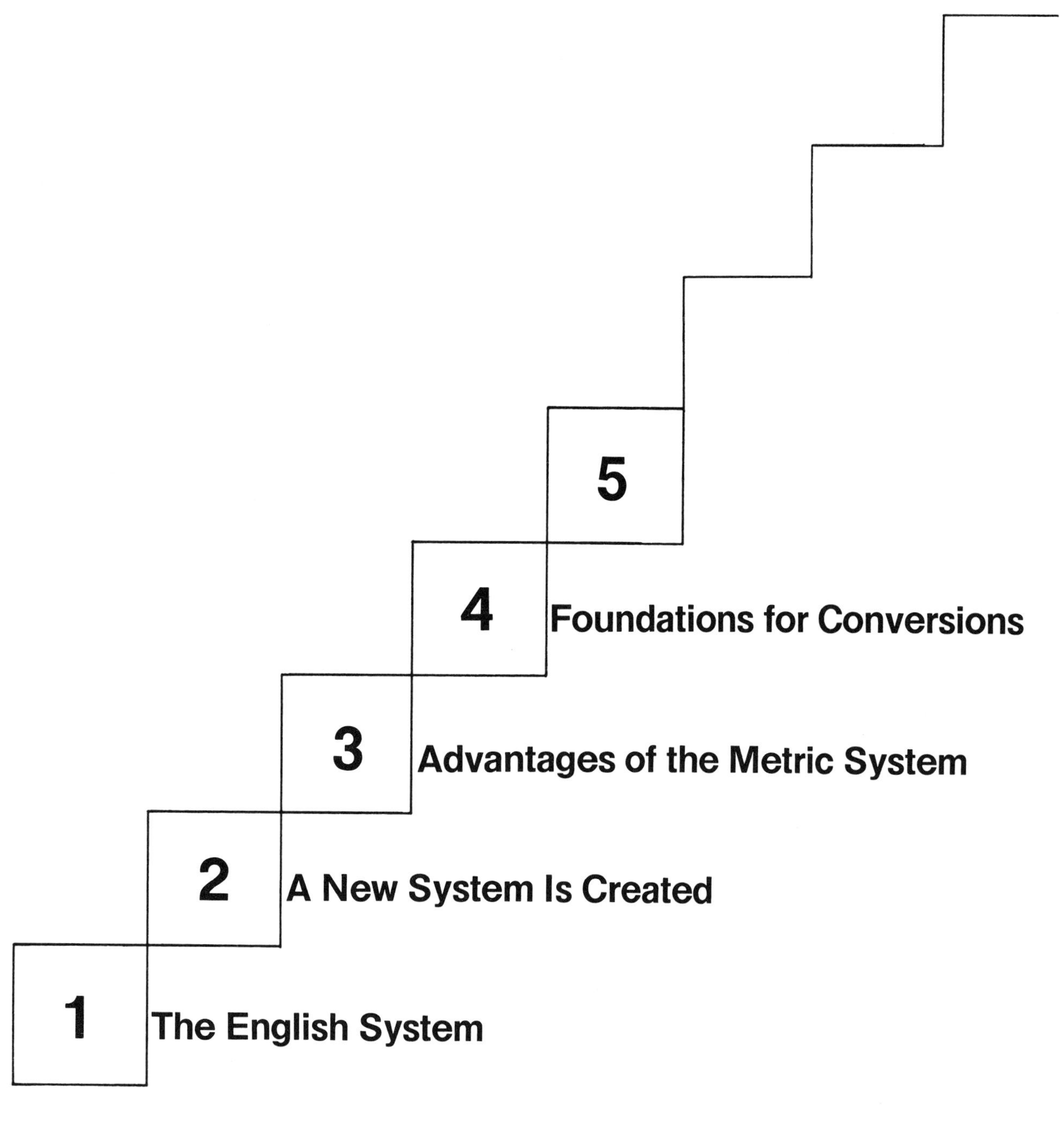
5
4
Foundations for Conversions
3
Advantages of the Metric System
2
A New System Is Created
1
The English System

Introduction

Tell the meaning or significance of each mark, sign, word, or symbol on this page. Then you'll be ready to start Unit 5.

numerator	kilometer
÷ ____________	dekameter
denominator	millimeter
mm	centimeter
cm	French Revolution
dm	French Academy of Science
km	National Assembly of France
deci	1791
centi	ratio
milli	conversion
prefix	

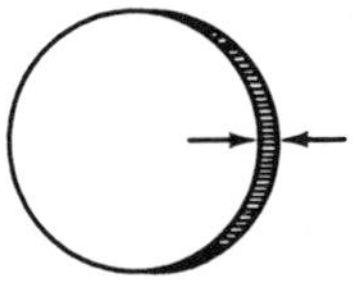

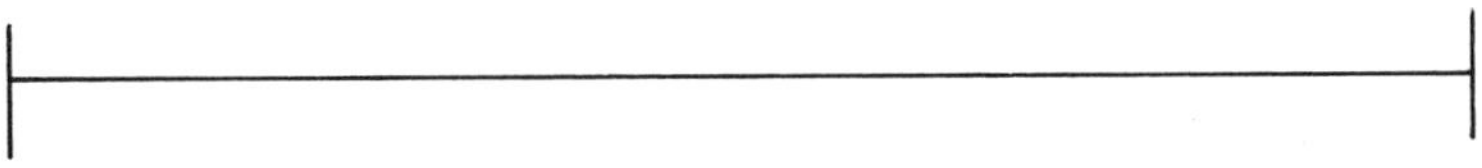

You will need to use more than one unit to finish the measuring job. There is more than one correct way, but you must decide which units would be most convenient to measure:

millimeter
centimeter
decimeter
meter
dekameter
kilometer

How long or how wide your pencil is ______________________________

How high or how far you can jump ______________________________

How far you can throw a baseball ______________________________

How far a racer could ride a bicycle in one day ____________________

How far you can shoot an arrow ______________________________

How long your thumbnail is ______________________________

How tall you are ______________________________

The width of the lead in your unsharpened pencil ____________________

Poll the class to determine which parents use metric measurement in their work.

Turn to page 54 for the answers to Unit 5, Frames 134-179.

134. When you measure something you must do two things: (1) you decide what kind of units of measurement you will use, inches, feet, centimeters, meters, or such, and (2) you count or measure to determine how many of your selected units are required to equal the item being measured.

For example, to measure your height you first decide what ____________ of measurement to use, then you count or ____________ how many are required to equal your height.

135. If a monkey is 82 cm tall, the unit of measurement is ____________ and the count or measure is ____________.

136. A nickel has a radius of 1 centimeter. The unit of this measurement is ____________ and the measure is ____________.

137. A dime is approximately 1 millimeter thick. The two parts of this measurement are the ____________ of the measurement and the m____________ stating how many.

138. What is the measure in centimeter units of a board 2 meters long?

139. The ratio of $\frac{1 \text{ m}}{100 \text{ cm}}$ represents a distance measured to be 1 meter divided by a distance measured to be 100 centimeters. Since 1 m and 100 cm are both the ____________ distance the ratio of $\frac{1 \text{ m}}{100 \text{ cm}}$ is equal to ____________.

140. $\frac{8}{1} \times \frac{1 \text{ m}}{100 \text{ cm}} = 8$ $\frac{8}{1} \times 1 = 8$

Both of the above multiplications produce an answer of 8. The answer eight is simply a measure. The answer contains no ____________ of measurement.

141. Which of the following is a measure, and not a measurement?

6 cm 14 km 4 2 m

142. Which of these ratios is the same as a measure of one: A, B, or C?

(A) $\frac{1\ m}{1\ cm}$ (B) $\frac{10\ m}{1\ km}$ (C) $\frac{1\ m}{100\ cm}$

143. A distance of 10 meters contains a number 10 called the ____________ of the distance. The word meter is the name of the ____________ of measurement.

144. If 5, which is only a measure of a distance, is multipled by $\frac{1000\ mm}{1\ m}$ the answer (product) is 5. Since $\frac{1000\ mm}{1\ m}$ really represents ____________ the answer 5, of course, can have no units attached to it.

145. $26 \times \frac{1\ km}{1000\ m}$ = ____________.

There is no ____________ of measurement in this product. The product consists only of a ____________.

146. $\frac{16}{1} \times \frac{1\ cm}{10\ mm}$ = ____________.

Again the product is only a ____________ and has no ____________ included in it.

147. A measure multiplied by a ratio representing 1 will produce a product consisting only of a ____________.

148. 4 km × 1 = 4 km

One kilometer given 4 times equals 4 kilometers. 4 km × $\frac{10\ dm}{1\ m}$ = ?

149. $\frac{12\ cm}{1} \times \frac{1\ m}{100\ cm}$ = ? (Remember $\frac{1\ m}{100\ cm}$ is the same as 1.)

150. $2 + 2 + 2 + 2 + 2 + 2 =$ ____________.

It can also be stated that 2 X ____________ = 12.

151. Multiplication is only a simplified form to show the longer process of ____________. 10 teeth added 10 times results in a total of ____________ teeth.

152. In the measurement, 10 teeth, the 10 is the ____________ and teeth is the name of the ____________.

153. 20 apples X 2, or 20 apples + 20 apples = ?

154. $4 \times 4 =$ ____________. 4 km X 4 = ____________.

$\frac{4 \text{ km}}{1} \times \frac{100 \text{ cm}}{1 \text{ m}} =$ ____________.

155. It was stated earlier that placing a value over the number 1 did not change the value. 3 X 1000 mm can be changed to $\frac{3}{1} \times \frac{1000 \text{ mm}}{1} =$ ____________ ____________.

156. $\frac{5}{1} \times \frac{100 \text{ cm}}{1} =$ ____________. $\frac{5 \text{ cm}}{1} \times \frac{100}{1} =$ ____________.

157. $\frac{2 \text{ cm}}{1} \times \frac{100}{1} = 200 \text{ cm}.$ $\frac{40 \text{ m}}{1} \times \frac{10}{1} =$ ____________.

158. $\frac{900 \text{ cm}}{1} \times \frac{1}{1} =$ ____________. $\frac{900}{1} \times \frac{1}{1} =$ ____________.

159. 900 cm can be called a measurement since it is a measure of 900 units called centimeters. 9000 is called a ____________ since it contains no units. 16 m is called a ____________.

160. 500 dm is a ____________. 5 is a ____________.

161. $\frac{4\text{ m}}{1} \times \frac{10}{1} = 40$ m. When multiplying a measurement by a measure, the product is a ____________.

$\frac{6\text{ cm}}{1} \times \frac{100\text{ cm}}{1\text{ m}} =$ ____________.

162. $\frac{4}{1} \times 1 = 4$. When multiplying a measure by a measure the result is a ____________.

$\frac{6}{1} \times \frac{10}{10} = 6$. $\frac{7}{1} \times \frac{1\text{ m}}{100\text{ cm}} =$ ____________.

163. A fraction in which the value of the numerator and the denominator are equal results in a measure of ____________.

$\frac{4}{4} = 1$. $\frac{5\text{ apples}}{5\text{ apples}} =$ ____________. $\frac{12\text{ eggs}}{1\text{ dozen eggs}} =$ ____________.

164. $\frac{1\text{ year}}{?\text{ months}} = 1$. $\frac{100\text{ cm}}{?\text{ m}} = 1$.

In both of the above cases dividing the denominator into the numerator should result in 1 which really means the lower value will divide into the upper value exactly 1 ____________.

165. $\frac{4}{2} = 2$. $\frac{10}{2} =$ ____________.

In both of these cases the fraction is equal to a measure which represents how many ____________ the denominator will divide into the numerator.

166. $\frac{10\text{ apples}}{2\text{ apples}} = 5$. $\frac{10\text{ meters}}{2\text{ meters}} =$ ____________.

In both cases the fraction is equal to a ____________ which represents how many ____________ the denominator will divide into the numerator.

167. $\frac{20\text{ m}}{5\text{ m}}$. This is the same as writing how many ____________ 5 meters can be divided into 20 meters. The answer is 4 ____________, *not* 4 meters.

168. $\frac{24\text{ eggs}}{1\text{ dozen eggs}}$ is the same as writing how many times ____________ eggs can be divided into 24 eggs. The answer is ____________ times.

169. $\frac{300\text{ cm}}{1\text{ m}}$ is the same as writing how many times ____________ can be divided into ____________. The answer is ____________.

170. Two of the following fractions are not equal to 1. Which fraction *is* equal to the measure 1: A, B, or C?

(A) $\frac{12\text{ inches}}{1\text{ foot}}$ (B) $\frac{12\text{ years}}{12\text{ months}}$ (C) $\frac{2\text{ m}}{10\text{ dm}}$

171. $\frac{1000\text{ mm}}{1\text{ m}} =$ ____________. The answer here is *not* a measurement, but a ____________. It means how many ____________ 1 m goes into 1000 mm.

172. $5 \times 1 =$ ____________.

$\frac{5\text{ m}}{1} \times \frac{1\text{ m}}{10\text{ dm}} =$ ____________.

$\frac{6\text{ cm}}{1} \times \frac{1\text{ m}}{100\text{ cm}} =$ ____________.

173. $\frac{120\text{ m}}{1} \times \frac{100\text{ cm}}{1\text{ m}} =$ ____________.

$\frac{15\text{ mm}}{1} \times \frac{1\text{ m}}{1000\text{ mm}} =$ ____________.

$\frac{7\text{ km}}{1} \times \frac{1000\text{ m}}{1\text{ km}} =$ ____________.

174. $\frac{4\text{ cm}}{1} \times \frac{1\text{ m}}{100\text{ cm}} =$ ____________.

It is clear that multiplying a value like 4 cm by a ratio equal to 1 does not change the value of the 4 cm. A measurement multiplied by a ____________ equal to 1 does not change the value of the given ____________.

175. (A) 10 (B) 10 cm

Which of the above is a unit number: A or B ?

Which of the above is a pure number: A or B ?

176. (A) $\frac{1\text{ m}}{100\text{ cm}}$ (B) $\frac{100\text{ km}}{1\text{ m}}$

Which ratio is equal to 1: A or B ?

177. $5 \times \frac{1000\text{ mm}}{1\text{ m}} =$ ____________. $10\text{ cm} \times \frac{10\text{ dm}}{1\text{ m}} =$ ____________.

$\frac{15\text{ km}}{1} \times \frac{1000\text{ m}}{1\text{ km}} =$ ____________.

178. Change 50 cm to a fraction. Change $\frac{12\text{ m}}{1}$ to a whole number.

179. What are two ratios showing the way centimeters and meters are related? ____________ and ____________.

END OF UNIT 5

ANSWERS

Unit 5 MECHANICS OF MEASUREMENT Frames 134-179

134. units; measure
135. centimeter (cm); 82
136. centimeter (cm); one (1)
137. units; measure
138. 200
139. same; one (1)
140. unit
141. 4
142. C
143. measure; unit
144. one (1)
145. 26; unit; measure
146. 16; measure; unit
147. measure
148. 4 km
149. 12 cm
150. 12; 6
151. addition (adding); 100
152. measure; unit
153. 40 apples
154. 16; 16 km; 4 km
155. 3000 mm
156. 500 cm; 500 cm
157. 400 m
158. 900 cm; 900
159. measure; measurement
160. measurement; measure
161. measurement; 6 cm
162. measure; 7
163. 1; 1; 1
164. 12; 1; time
165. 5; times
166. 5; measure; times
167. times; times
168. 1 dozen; 2
169. 1 m; 300 cm; 3 times
170. A
171. one (1); measure; times
172. 5; 5 m; 6 cm
173. 120 m; 15 mm; 7 km
174. 4 cm; ratio; measurement
175. B; A
176. A
177. 5; 10 cm; 15 km
178. $\frac{50\text{cm}}{1}$; 12 m
179. $\frac{1\text{ m}}{100\text{ cm}}$; $\frac{100\text{ cm}}{1\text{ m}}$

Unit 6

LINEAR CONVERSION SKILLS

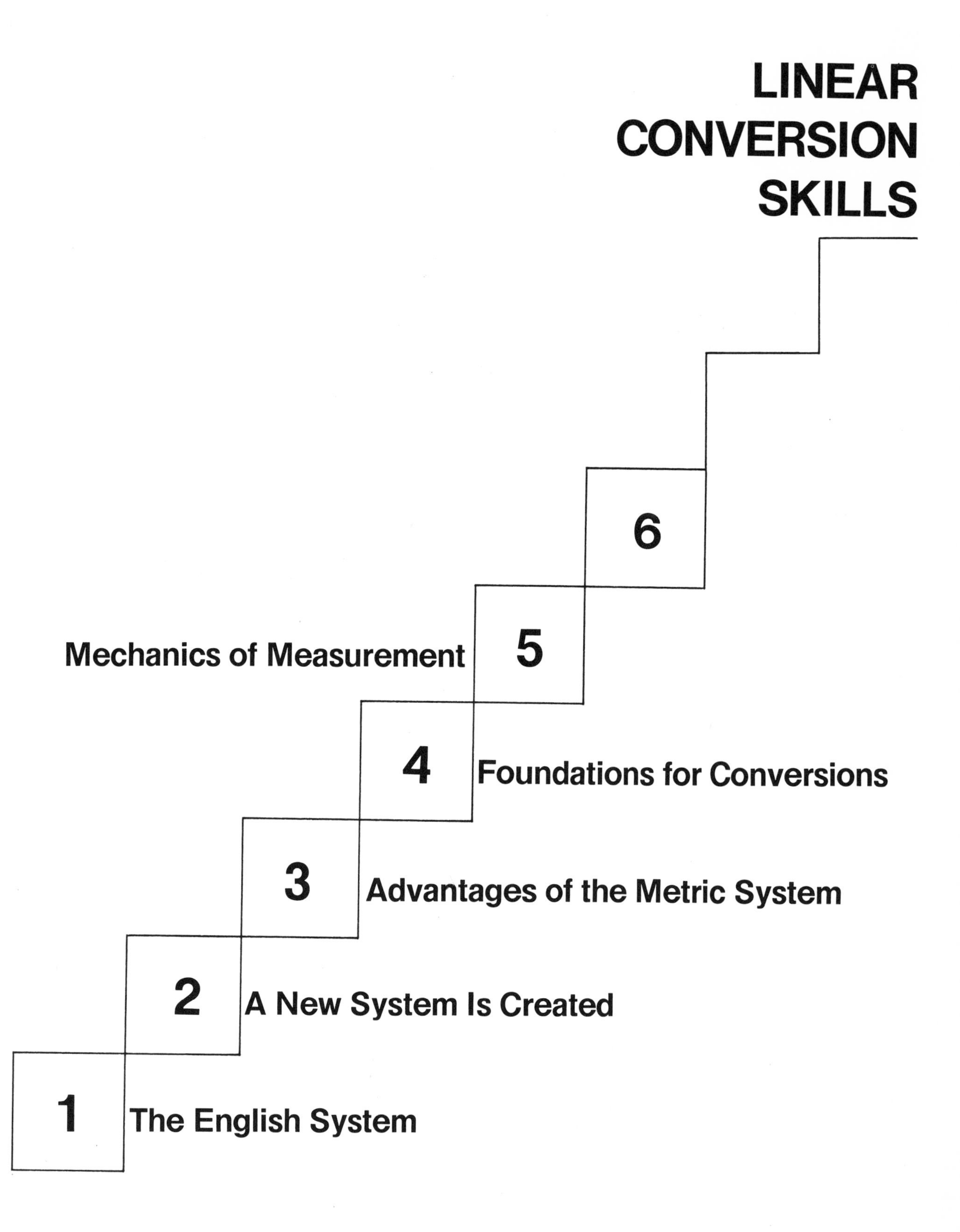

Introduction

Think Metrically !

Without measuring first, guess approximate metric distances of:

____________ diameter of a nickel

____________ thickness of a dime

____________ diameter of a half-dollar

____________ distance from a gym floor to a basketball rim

____________ diameter of a golfball

____________ diameter of an l.p. record

____________ distance between the goal posts on opposite ends of a football field

____________ length of a regulation football

____________ holes in a salt shaker

____________ length and width of this page

____________ height of a flagpole in front of a school

____________ width of the door leading into the room you are in

____________ length of your shoe

____________ height of the Statue of Liberty

____________ circumference of a softball

Units in the metric system convert easily from one unit to another. This unit shows you how easily metric units convert.

Big problems are easily solved – too good to be true? TRY IT!

VOLUME CONVERSIONS

1 cubic inch = 16 cubic centimeters	1 cubic centimeter = 0.06 cubic inch
1 cubic foot = 0.03 cubic meters	1 cubic meter = 35 cubic feet
1 cubic yard = 0.8 cubic meters	1 cubic meter = 1.3 cubic yards
1 gallon = 0.004 centimeters	1 cubic meter = 250 gallons

Turn to page 66 for the answers to Unit 6, Frames 180-243.

180. The measurement of a distance expressed in one type of unit can be easily converted to the same measurement expressed in another type of unit by a process called conversion. Five meters can be con_________ed to 500 centimeters by a process called _________version.

181. You will learn the formal method of converting a measurement in one kind of unit to another kind of unit. The ____________ method of __________sion is the method used by scientists and technicians.

182. This ____________ method is the way __________sions are shown when writing out calculations in science and mathematics. Short cut methods are possible for converting mentally from one unit to another, but only the ____________ method is acceptable for written work.

183. The first con__________ is to change 4 km to meters. The relationship of *km* and *m* units can be expressed as either ____________ or ____________.

184. You have learned that a value of a distance can be multiplied by the number ____________ without changing the value of the distance. You have also learned that a distance such as 4 km can be multiplied by a ratio that is the same as ____________ without changing the value of the 4 km.

185. In the conversion of 4 km to meter units, change the 4 km to a fraction and multiply it by the ratio of the two units involved, km and m. $\frac{4 \text{ km}}{1} \times \frac{1 \text{ km}}{1000 \text{ m}}$ is one way the problem can be written and $\frac{4 \text{ km}}{1} \times \frac{?}{?}$ is the other way it can be written.

186. In the second part of each example, $\frac{4 \text{ km}}{1} \times \frac{1 \text{ km}}{1000 \text{ m}}$ and $\frac{4 \text{ km}}{1} \times \frac{1000 \text{ m}}{1 \text{ km}}$, the ratios have the same value as the number ____________ so 4 km X 1 does not change the *value* of the 4 km. Notice what happens if the multiplication is done this way:

$$\frac{\overset{4}{\cancel{4 \text{ km}}}}{1} \times \frac{1000 \text{ m}}{\underset{1}{\cancel{1 \text{ km}}}} = \frac{4000 \text{ m}}{1} \text{ or } \_\_\_\_\_\_\_\_\_\_\_\_.$$

187. You have probably "cancelled" or simplified like this before. For example, $\frac{\overset{1}{\cancel{3}}}{1} \times \frac{4}{\underset{1}{\cancel{3}}} = 4$ and

$\frac{\overset{4}{\cancel{40}}}{1} \times \frac{1000}{\underset{1}{\cancel{10}}} =$ ____________. Copy this problem and solve it by simplifying.

$\frac{16}{1} \times \frac{3}{4} =$ ____________.

188. $\frac{\overset{4}{\cancel{4\text{ km}}}}{1} \times \frac{1000\text{ m}}{\underset{1}{\cancel{1\text{ km}}}}$

To remove the km units in this problem proceed as follows: 1 km divides evenly into both 4 km and 1 km. 1 km divides into 4 km ____________ times. 1 km divides into 1 km ____________ time.

189. $\frac{8\text{ km}}{1} \times \frac{1000\text{ m}}{1\text{ km}}$

To simplify these ratios you divide 1 km into both ____________ and ____________.

190. $\frac{6\text{ m}}{1} \times \frac{10\text{ dm}}{1\text{ m}} =$

Copy this multiplication and simplify before you multiply (by "cancelling").

191. Simplify this problem. $\frac{8\text{ cm}}{1} \times \frac{10\text{ mm}}{1\text{ cm}} =$ ____________.

192. Simplify this conversion once more for practice: $\frac{20\text{ dm}}{1} \times \frac{10\text{ cm}}{1\text{ dm}} =$ ____________.

193. Multiplying a given measurement stated in kilometers by a ratio that has the same value as ____________ will not change the value of the given measurement. If you do not simplify, then

$\frac{4\text{ km}}{1} \times \frac{1000\text{ m}}{1\text{ km}} = 4\text{ km}$ because $\frac{1000\text{ m}}{1\text{ km}}$ is the same as 1. If you *do* simplify, then $\frac{\overset{4}{\cancel{4\text{ km}}}}{1} \times \frac{1000\text{ m}}{\underset{1}{\cancel{1\text{ km}}}} =$

4000 m. Are the two answers, 4 km and 4000 m, really the same distance? (*Yes or no.*)

194. When $\frac{4 \text{ km}}{1} \times \frac{1000 \text{ m}}{1 \text{ km}}$ *is not* simplified the answer is ____________ km. When the problem is simplified $\left(\frac{\overset{4}{\cancel{4 \text{ km}}}}{1} \times \frac{1000 \text{ m}}{\underset{1}{\cancel{1 \text{ km}}}}\right)$ the answer is ____________ m.

195. Try another conversion: 16 m changed to mm units. The conversion would first be written as $\frac{16 \text{ m}}{1} \times \frac{1000 \text{ mm}}{1 \text{ m}}$. Multiplying 16 m by a ratio equal to 1 will produce an answer in ____________ units *unless* the 16 m and the 1 m are "cancelled" out by dividing each by ____________.

196. $\frac{16 \text{ m}}{1} \times \frac{1000 \text{ mm}}{1 \text{ m}}$. To simplify the 16 m and the 1 m, each should be divided by 1 m. 1m divided into 1 m is ____________ and 1 m divided into 16 m is ____________.

197. When we simplify, the meter units are removed $\frac{\overset{16}{\cancel{16 \text{ m}}}}{1} \times \frac{1000 \text{ mm}}{\underset{1}{\cancel{1 \text{ m}}}}$ and the answer produced is the same distance as 16 m, but is stated as 16 000 ____________.

198. $\frac{\overset{16}{\cancel{16 \text{ m}}}}{1} \times \frac{1000 \text{ mm}}{\underset{1}{\cancel{1 \text{ m}}}} =$ ____________.

199. Change 500 cm to meters. $\frac{\overset{5}{\cancel{500 \text{ cm}}}}{1} \times \frac{1 \text{ m}}{\underset{1}{\cancel{100 \text{ cm}}}} = \frac{5 \text{ m}}{1}$ or ____________.

200. Convert 16 000 m to km: ____________ $\times \frac{1 \text{ km}}{1000 \text{ m}} =$ ____________.

201. $\frac{\overset{16}{\cancel{16\,000\text{ m}}}}{1} \times \frac{1\text{ km}}{\underset{1}{\cancel{1000\text{ m}}}} = \frac{16\text{ km}}{1}$ or 16 km

What unit number was divided into both 16 000 m and 1000 m to simplify them to 16 and 1? ____________ ____________.

202. Convert 40 000 cm to meters. $\frac{40\,000\text{ cm}}{1} \times \frac{1\text{ m}}{100\text{ cm}} =$ ____________. Show the full conversion on your answer sheet.

203. To convert 40 000 cm to meters multiply by the ratio of meters to centimeters in which the denominator of the ratio is not 1 m but ____________. This will allow simplification and removal of the unit ____________.

204. In converting 5 meters to centimeters the distance is given in ____________ units which are to be changed to the desired units of ____________. The ratio of the two units to each other is either $\frac{100\text{ cm}}{1\text{ m}}$ or ____________.

205. In converting 5 m to centimeters, which ratio, A or B, will make possible the removal of the meter units? $\frac{5\text{ m}}{1} \times \left(\text{(A)}\ \frac{1\text{ m}}{100\text{ cm}} \text{ or (B) } \frac{100\text{ cm}}{1\text{ m}}\right)$.

206. When changing a measurement described in meters to the same measurement given in centimeters, you must multiply by a ratio of ____________ to ____________. The ratio must be stated in this form since the ____________ units must be "cancelled" out, leaving only the centimeter units.

207. Convert 9 m to centimeters. $\frac{9\text{ m}}{1} \times \left(\text{(A)}\ \frac{100\text{ cm}}{1\text{ m}} \text{ or (B) } \frac{1\text{ m}}{100\text{ cm}}\right) = 900\text{ cm}.$

Which form of the ratio should be used, A or B?

208. Convert 40 m to millimeters. $\frac{40\text{ m}}{1} \times \left((A)\ \frac{1\text{ m}}{1000\text{ mm}} \text{ or } (B)\ \frac{1000\text{ mm}}{1\text{ m}}\right) = 40\,000\text{ mm}.$

Which ratio should be used, A or B?

209. Convert 50 km to meters. $\frac{50\text{ km}}{1} \times \left((A)\ \frac{1000\text{ m}}{1\text{ km}} \text{ or } (B)\ \frac{1\text{ km}}{1000\text{ m}}\right) = 50\,000\text{ m}.$

Which ratio should be used, A or B?

210. Change 4 km to meters by the formal method of conversion. Show your work.

211. Change 6 km to meters by formal conversion. Show your work.

212. Change 6000 m to millimeters by the formal method of conversion.

213. Change 10 m to millimeters.

214. 400 cm equals ____________ meters. Show this by formal conversion.

215. Change 40 cm to meters. In *this* conversion a new situation arises: $\frac{40\text{ cm}}{1} \times \frac{1\text{ m}}{100\text{ cm}}.$
The question is what will divide into both 40 cm and 100 cm evenly. The largest unit number that will divide evenly into each is 10 cm. Write down the conversion and simplify by dividing each by 10 cm. *Do not complete the conversion beyond simplifying.*

216. $\frac{\overset{4}{\cancel{40\text{ cm}}}}{1} \times \frac{1\text{ m}}{\underset{10}{\cancel{100\text{ cm}}}} = \frac{4\text{ m}}{10}$

It is *not* good form to let the answer remain as a fraction. Change *4 m* to a unit that is in decimal form.

Hint: $10\overline{\smash{)}4.0\text{ m}}$ with quotient $.?\text{ m}$

217. $\frac{4}{10}$ is written "four tenths" or as .4,

$\frac{4\text{ m}}{10}$ is written as .4 m,

$\frac{4\text{ m}}{100}$ is written as .04 m,

$\frac{4\text{ m}}{1000}$ is written as ______________.

218. $\frac{40}{100}$ is written as .40.

$\frac{40}{100}$ can be simplified to $\frac{4}{10}$ and .40 simplified to ______________.

$\frac{47}{100}$ cannot be simplified and is written as ______________.

219. $\frac{8}{1} = 8.$ $\frac{8}{10} = .8$

$\frac{8}{100} =$.______________. $\frac{8}{1000} =$.______________.

$\frac{8}{10\,000} =$.______________.

220. $\frac{13}{1} =$ ______________; $\frac{13}{10} =$ ______.______; $\frac{13}{100} =$.______________.

221. $\frac{5}{10} =$.______________; $\frac{71}{10} =$ ______.______; $\frac{77}{100} =$.______________.

222. $\frac{11}{100} =$.______________; $\frac{125}{10} =$ ______.______; $\frac{7894}{100} =$ ______.______.

223. Convert 30 mm to meters by formal conversion. $\frac{30\text{ mm}}{1} \times \frac{1\text{ m}}{1000\text{ mm}} =$ ______*______.

*Hint: Simplify by dividing by 10 mm.

224. Change 43 mm to meters. $\frac{43 \text{ mm}}{1}$ X ______*______ = ____________.

*Hint: 43 mm and 1000 mm *must* be "cancelled" to remove the millimeter units. If nothing seems to divide evenly into each, then divide each by 1 mm.

225. It is a mathematical law that multiplying a value by a r____________ that is the same as ____________ will not change the value.

226. The formal system can be used to make any type of conversion in addition to conversion of metric units. The *approximate* relationship of inches to meters is $\frac{40 \text{ inches}}{1 \text{ meter}}$. Change 4 m to inches.

$\frac{4 \text{ m}}{1} \times \frac{40 \text{ inches}}{1 \text{ m}} =$ ____________. Show the work of the formal conversion.

227. By formal conversion change 12 quarts to gallons. *Show the simplifying.*

$\frac{12 \text{ quarts}}{1}$ X ____________ = ____________. (There are 4 quarts in 1 gallon.)

228. Convert 4 yards to inches by formal conversion. (1 yard is 36 inches.)

229. Sometimes a conversion must be made in two steps, for example, change 4 km to millimeters. The *first* step is to change 4 km to meter units. The *second* step is then to change the meter units to millimeter units. *Both* steps can be written into *one* conversion.

$\frac{4 \text{ km}}{1} \times \frac{1000 \text{ m}}{1 \text{ km}} \times \frac{1000 \text{ mm}}{1 \text{ m}} =$ ____________. Copy this conversion and "cancel" out the km and m units. Complete the conversion to get millimeters.

230. Change 12 km to *centi*meters by putting both ratios in one formal conversion. (Hint: a meter-kilometer ratio first and then a centimeter ratio.)

231. Convert 4 km to inches. *Hint:* First multiply 4 km by a ratio of meters to kilometers that will cause the km to be removed. Then multiply by the ratio of inches to meters to remove the meters. Show your work of simplifying.

232. Change 10 years to hours. *Hint:* $\frac{10 \text{ years}}{1} \times \frac{365 \text{ days}}{1 \text{ year}} \times \frac{24 \text{ hours}}{1 \text{ day}} =$ ____________. Show all work.

233. The advantage of the ____________ method of converting a value from one unit to another is that you show in a simple manner how you make the conversion. Change 5 pounds to ounces. (The rationship of the two units is $\frac{16 \text{ ounces}}{1 \text{ pound}}$ or $\frac{1 \text{ pound}}{16 \text{ ounces}}$.)

234. Convert 3600 seconds to hours, "cancel" as far as possible.

$\frac{3600 \text{ seconds}}{1} \times \frac{1 \text{ minute}}{60 \text{ seconds}} \times \frac{1 \text{ hour}}{60 \text{ minutes}} =$ ____________.

235. You have learned how to convert units by the ____________ method of conversion. This method of ____________ is required when showing in writing how one unit has been converted to another.

236. The advantage of the formal method of ____________ is that it shows in a simple manner the mathematics used in converting a measurement in one type of units to another type. The law that this method is based upon states that a number multiplied by a ____________ which is the same as ____________ does not change its value.

237. The value of a distance is not changed in formal conversions but the ____________ are changed from the *given* units to the *desired* units. If this is done correctly, the value of the given units will ____________ the value of the desired units.

238. Change 98 days to weeks. $\frac{98 \text{ days}}{1} \times$ ____________ = ____________.

239. Convert 10 weeks to hours.

240. Change 2 miles to inches. $\frac{2 \text{ miles}}{1} \times \frac{5280 \text{ feet}}{1 \text{ mile}} \times \frac{?}{?} =$ ____________.

241. Change 30 mm to centimeters. (There are 10 mm per 1 cm.)

242. Change 440 dm to inches.

243. Convert 15 000 cm to km.

END OF UNIT 6

ANSWERS

Unit 6 LINEAR CONVERSION SKILLS Frames 180-243

180. converted; conversion

181. formal; conversion

182. formal; conversions; formal

183. conversion; $\frac{1\text{ km}}{1000\text{ m}}$; $\frac{1000\text{ m}}{1\text{ km}}$

184. 1; 1

185. $\frac{1000\text{ m}}{1\text{ km}}$

186. one (1); 4000 m

187. 4000; $\frac{\overset{4}{\cancel{16}}}{1} \times \frac{3}{\underset{1}{\cancel{4}}} = \frac{12}{1}$ or 12

188. four (4); one (1)

189. 8 km; 1 km (either order)

190. $\frac{\overset{6}{\cancel{6\text{ m}}}}{1} \times \frac{10\text{ dm}}{\underset{1}{\cancel{1\text{ m}}}} = 60\text{ dm}$

191. $\frac{\overset{8}{\cancel{8\text{ cm}}}}{1} \times \frac{10\text{ mm}}{\underset{1}{\cancel{1\text{ cm}}}} = 80\text{ mm}$

192. $\frac{\overset{20}{\cancel{20\text{ dm}}}}{1} \times \frac{10\text{ cm}}{\underset{1}{\cancel{1\text{ dm}}}} = 200\text{ cm}$

193. one (1); yes

194. 4 km; 4000 m

195. meter; 1 meter

196. 1; 16

197. millimeters (mm)

198. 16 000 mm

199. 5 m

200. $\frac{16\,000\text{ m}}{1}$; 16 km

201. 1000 m

202. $\frac{\overset{400}{\cancel{40\,000\text{ cm}}}}{1} \times \frac{1\text{ m}}{\underset{1}{\cancel{100\text{ cm}}}} = 400\text{ m}$

203. 1 hundred cm; cm

204. meter; centimeters; $\frac{1\text{ m}}{100\text{ cm}}$

205. B

206. 100 centimeters; 1 meter; meter

207. A

208. B

209. A

210. $\frac{\overset{4}{\cancel{4\text{ km}}}}{1} \times \frac{1000\text{ m}}{\underset{1}{\cancel{1\text{ km}}}} = 4000\text{ m}$

211. $\frac{\overset{6}{\cancel{6\text{ km}}}}{1} \times \frac{1000\text{ m}}{\underset{1}{\cancel{1\text{ km}}}} = 6000\text{ m}$

212. $\frac{\overset{6000}{\cancel{6000\text{ m}}}}{1} \times \frac{1000\text{ mm}}{\underset{1}{\cancel{1\text{ m}}}} = 6\,000\,000\text{ mm}$

213. $\frac{\overset{10}{\cancel{10\text{ m}}}}{1} \times \frac{1000\text{ mm}}{\underset{1}{\cancel{1\text{ m}}}} = 10\,000\text{ mm}$

214. $\frac{\overset{4}{\cancel{400\text{ cm}}}}{1} \times \frac{1\text{ m}}{\underset{1}{\cancel{100\text{ cm}}}} = 4\text{ m}$

215. $\frac{\overset{4}{\cancel{40\text{ cm}}}}{1} \times \frac{1\text{ m}}{\underset{10}{\cancel{100\text{ cm}}}}$

216. $10\overline{)4.0\text{ m}}$ = .4 m

217. .004 m

218. .4$\not{0}$ (or .4); .47

219. .08; .008; .0008

220. 13; 1.3; .13

221. .5; 7.1; .77

222. .11; 12.5; 78.94

223. $\frac{\overset{3}{\cancel{30\text{ mm}}}}{1} \times \frac{1\text{ m}}{\underset{100}{\cancel{1000\text{ mm}}}} = \frac{3\text{ m}}{100}$ or .03 m

224. $\frac{\overset{43}{\cancel{43\text{ mm}}}}{1} \times \frac{1\text{ m}}{\underset{1000}{\cancel{1000\text{ mm}}}} = \frac{43\text{ m}}{1000}$ or .043 m

225. <u>ratio</u>; one (1)

226. $\frac{\overset{4}{\cancel{4\text{ m}}}}{1} \times \frac{40\text{ inches}}{\underset{1}{\cancel{1\text{ m}}}} = 160\text{ inches}$

227. $\frac{\overset{3}{\cancel{12\text{ quarts}}}}{1} \times \frac{1\text{ gal.}}{\underset{1}{\cancel{4\text{ qts.}}}} = 3\text{ gal.}$

228. $\frac{\overset{4}{\cancel{4\text{ yards}}}}{1} \times \frac{36\text{ inches}}{\underset{1}{\cancel{1\text{ yard}}}} = 144\text{ inches}$

229. $\frac{\overset{4}{\cancel{4\text{ km}}}}{1} \times \frac{\overset{1000}{\cancel{1000\text{ m}}}}{\underset{1}{\cancel{1\text{ km}}}} \times \frac{1000\text{ mm}}{\underset{1}{\cancel{1\text{ m}}}} = 4\,000\,000\text{ mm}$

230. $\frac{\overset{12}{\cancel{12\text{ km}}}}{1} \times \frac{\overset{1000}{\cancel{1000\text{ m}}}}{\underset{1}{\cancel{1\text{ km}}}} \times \frac{100\text{ cm}}{\underset{1}{\cancel{1\text{ m}}}} =$

1 200 000 cm

231. $\frac{\overset{4}{\cancel{4\text{ km}}}}{1} \times \frac{\overset{1000}{\cancel{1000\text{ m}}}}{\underset{1}{\cancel{1\text{ km}}}} \times \frac{40\text{ inches}}{\underset{1}{\cancel{1\text{ m}}}} =$

160 000 inches

232. $\frac{\overset{10}{\cancel{10\text{ years}}}}{1} \times \frac{\overset{365}{\cancel{365\text{ days}}}}{\underset{1}{\cancel{1\text{ year}}}} \times \frac{24\text{ hrs.}}{\underset{1}{\cancel{1\text{ day}}}} =$

87 600 hours

233. formal; $\frac{\overset{5}{\cancel{5\text{ pounds}}}}{1} \times \frac{16\text{ ounces}}{\underset{1}{\cancel{1\text{ pound}}}} = 80\text{ ounces}$

234. $\frac{\overset{\cancel{60}}{\cancel{3600\text{ seconds}}}}{1} \times \frac{\overset{1}{\cancel{1\text{ min.}}}}{\underset{1}{\cancel{60\text{ sec.}}}} \times \frac{1\text{ hr.}}{\underset{\underset{1}{\cancel{60}}}{\cancel{60\text{ min.}}}} =$

1 hour

235. formal; conversion

236. conversion; ratio; 1

237. units; equal

238.
$$\frac{\overset{14}{\cancel{98\text{ days}}}}{1} \times \frac{1\text{ week}}{\underset{1}{\cancel{7\text{ days}}}} = 14\text{ weeks}$$

239.
$$\frac{\overset{10}{\cancel{10\text{ wks.}}}}{1} \times \frac{\overset{7}{\cancel{7\text{ days}}}}{\underset{1}{\cancel{1\text{ wk.}}}} \times \frac{24\text{ hrs.}}{\underset{1}{\cancel{1\text{ day}}}} = 1680\text{ hours}$$

240.
$$\frac{\overset{2}{\cancel{2\text{ mi.}}}}{1} \times \frac{\overset{5280}{\cancel{5280\text{ ft.}}}}{\underset{1}{\cancel{1\text{ mile}}}} \times \frac{12\text{ in.}}{\underset{1}{\cancel{1\text{ ft.}}}} =$$

126 720 inches

241.
$$\frac{\overset{3}{\cancel{30\text{ mm}}}}{1} \times \frac{1\text{ cm}}{\underset{1}{\cancel{10\text{ mm}}}} = 3\text{ cm}$$

242.
$$\frac{\overset{44}{\cancel{440\text{ dm}}}}{1} \times \frac{\overset{1}{\cancel{1\text{ m}}}}{\underset{1}{\cancel{10\text{ dm}}}} \times \frac{40\text{ in.}}{\underset{1}{\cancel{1\text{ m}}}} = 1760\text{ in.}$$

alternate method:

$$\frac{\overset{440}{\cancel{440\text{ dm}}}}{1} \times \frac{4\text{ in.}}{\underset{1}{\cancel{1\text{ dm}}}} = 1760\text{ inches}$$

243.
$$\frac{\overset{15}{\overset{\cancel{150}}{\cancel{15\,000\text{ cm}}}}}{1} \times \frac{\overset{1}{\cancel{1\text{ m}}}}{\underset{1}{\cancel{100\text{ cm}}}} \times \frac{1\text{ km}}{\underset{100}{\underset{\cancel{1000}}{\cancel{1000\text{ m}}}}} =$$

$$\frac{15\text{ km}}{100} \text{ or } .15\text{ km}$$

Unit 7

DIMENSIONS OF MASS AND VOLUME

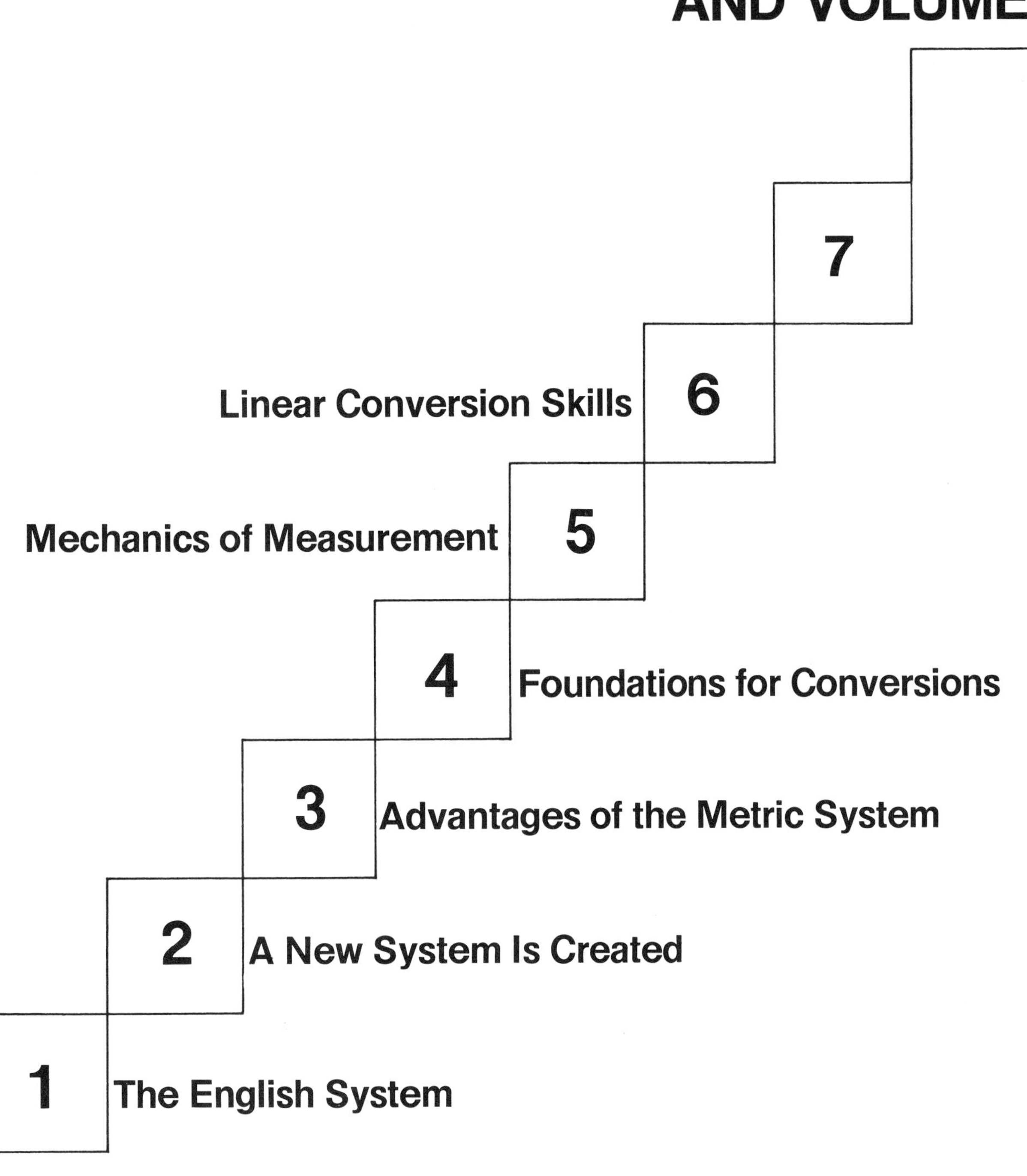

Introduction

How did you arrive to this page?

If you came through rather than around the frames in Unit 6 you've done well! You have solved many problems of the day – you have mastered the fundamentals of linear measurement in the metric system. Oh, yes, there's still the "micron" which you may want to look up, but now that you've gotten the rules down pat it should be easier sailing from now on.

Remember years ago when you were stalled by problems like this one? $\frac{3}{16}$ inch + $\frac{5}{9}$ yard + $\frac{3}{8}$ foot = ____________?

Making good friends with the metric system will improve your work and reduce frustration.

You will be pleased and relieved to know that volume, mass, and temperature in metric measurement are as well organized and as simple as length or linear metric measurements.

WEIGHT CONVERSIONS

U.S. to Metric	*Metric to U.S.*
1 ounce = 28 grams	1 gram = 0.035 ounce
1 pound = 0.45 kilogram	1 kilogram = 2.2 pounds
1 quart = 0.9 liter	1 liter = 1.06 quarts
1 ton = 0.9 metric ton	1 metric ton = 1.1 tons

Note: The symbol for centimeter is cm. A *square* centimeter is cm^2.
The only correct symbol for *cubic* centimeter (formerly cc) is cm^3.

Turn to page 82 for the answers to Unit 7, Frames 244-320.

244. Two advantages the metric system has over the English system for measuring distance are:
(1) All of the larger and smaller units of measurement are related to the meter and to each other by some multiple or division of ____________.
(2) The name for each unit contains a prefix that gives a ____________ indicating the size of the unit.

245. The pound is the basic unit of the English system for measuring weight. A metal weight similar to the standard pound weight will be pulled toward the earth by the earth's gravity with a force of one ____________.

246. A measurement of weight is a measurement of the force of gravitational attraction between the earth and a given material. The gravitational attraction between a pound of material and the ____________ will be a f____________ of 1 ____________.

247. If an object is attracted by gravity to the earth with a force that is twice as great as that of a one pound weight, the object is attracted by ____________ with a ____________ of 2 ____________.

248. The English pound is a unit of weight measurement which compares with the force of gravity between the ____________ and an object to be weighed. Anything that is attracted to the earth by the same amount of ____________ as that between the earth and a standard pound weight is said to have the ____________ of one pound.

249. If you own a pound of gold, then wherever you take this small fortune it should be attracted to the earth with a ____________ of ____________ ____________.

250. With the device shown on page 72 you could check to see if your gold weighed 1 pound. The procedure would be to:

(1) Place the standard 1 pound weight on the spring scale and note the distance the 1 pound weight was pulled downward by gravity.
(2) Next, the gold would be placed alone on the pan and if the spring stretched downward the same ____________ on the scale, the gold weighed one ____________.

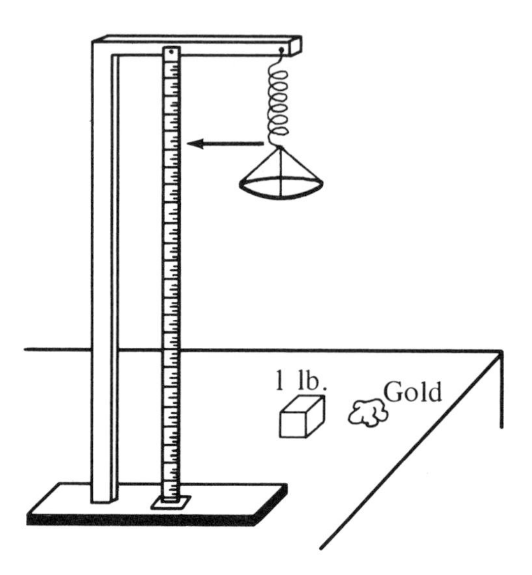

251. A basic weakness in the English pound unit is evident. If we took our gold nugget on an orbital ride around the earth and checked with the spring scale while in a "weightless" condition, the spring scale would read ____________ pounds. The amount of the gold in the nugget would still be the same but its ____________ would seem to change from 1 to 0 pound units.

252. The weakness of the English pound unit of measurement is due to the following: you do not measure the amount of a given material when you weigh it in pound units. You are actually measuring the ____________ of gravity on the material. It is possible in many situations for the measured force of ____________ between the earth and a given material to show considerable variation.

253. Very few people will lose their fortunes by having to sell their gold while in orbital flight. But if you are interested in exact measurements, there are situations on (and off) the earth where the force of the earth's ____________ varies considerably.

254. There are instruments that will measure the decrease in the force of ____________ as an object is raised away from the earth only a few millimeters. As one goes from a low spot on the earth, like Death Valley which is below sea level, to high mountainous areas, the decrease in the ____________ of gravity is important.

255. If you could buy gold by the pound, and if the seller were foolish enough to use a spring scale to measure its weight, where would you rather buy a pound of gold: at the top of Pikes Peak, or at a store in Death Valley?

256. The pound used for measuring units of ____________ was originally intended to measure amounts of material. We now know that a measurement of weight can change from place to place. To use the force of ____________ as a standard measuring stick is somewhat like using a length of elastic rubber for a standard unit to measure length.

257. The metric system unit for measuring the amount of matter in a material, or more precisely for measuring "mass," is called a gram. The ____________ is a unit for measuring ____________.

258. The gram unit (symbol g) is a standard unit for the measurement of ____________. The gram is also directly related to the *meter* unit, a measurement of ____________.

259. If a small metal container 1 cm wide by 1 cm long by 1 cm high is filled with pure water, at 4°C. the mass of the water in the container will be 1 gram. In the metric system the unit for measuring amount of material or mass is based on the distance measurement of ____________ ____________.

260. Water expands and contracts as the temperature changes. A cube, 1 cm by 1 cm by 1 cm, full of water at 80°F. would not be full if the water were cooled to 50°F. As warm water is cooled it c____________ until it reaches 4°C. or 39°F.

261. At 4°C. (or 39°F.) water has contracted to its smallest volume. If water is cooled below 39°F. or ____________, it expands in size until it becomes ice. Water in the liquid state has shrunk to its smallest volume at ____________°C.

262. Since water at a temperature of ____________ cannot contract or shrink any further, scientists once used a cubic centimeter of pure water at 4°C. as the standard mass of 1 ____________.

263. The standard metric unit for the measurement of mass was defined as a cube with each side ____________ in length of pure ____________ at a temperature of ____________.

264. A cube 1 centimeter on each side is called 1 cubic centimeter, or 1 cm^3. A gram of pure water at 4°C. would have a volume of 1 ____________ ____________ abbreviated* as 1 ____________.

* Symbols for units and prefixes are *symbols* and not abbreviations, though many people refer to them as abbreviations.

265. A m____________ of 1 gram, or 1 cm^3 of pure water at 4°C., is a relatively small amount of material. A large green garden pea takes up approximately 1 cm^3 of space. If a garden pea were made of water, it would have a mass of about ____________.

266. If three containers, each having a volume of 1 cm^3, were filled with lead, water, and cork, which container would have the greatest weight? ____________.The greatest mass? ____________.

267. If a cube, with each side one meter long (1 cubic meter), were first filled with air, then with cork, and finally with water, which material would have the greatest mass? ____________. Which the least mass? ____________.

268. One gram of a gas will fill 100 cm^3 of space, 1 gram of cork will fill 20 cm^3 of space and 1 gram of water will fill ____________ of space. Volume for volume, the mass of the gas is (more–less) than that of cork or water?____________.

269. The human body has about the same mass per unit of space as water. 1000 g of water takes up 1000 cm^3 of space. 1000 g of human body should, therefore, take up approximately ____________ cm^3 of space. A person with a body mass of 75 000 g would take up ____________ cm^3 of space.

270. Since the gram unit is so small, the term kilogram is used to name a mass containing ____________ g. One kilogram of water would contain 1000 grams of water which would have a volume of ____________ cm^3.

271. Since 1 kilogram (kg) of material normally *weighs* about 2.2 lb., we can think of 1 kg as about the mass of ____________ lb. of hamburger. Another easy way to estimate 1 kg is to think of the weight of a quart of water. One quart contains about 1000 cm^3. Therefore 1 qt. of water weighs about ____________ kg or ____________ lb.

272. If 1 kg is nearly equivalent to the amount of material or mass in ____________ lbs., then a person with a body mass of 50 kg would weigh about ____*____ lbs.

*Hint: $\frac{50\ \not{kg}}{1} \times \frac{2\text{ lb.}}{1\ \not{kg}} = ?$ lbs.

273. If a small foreign car has a mass of 1000 kg, then it weighs about ____*____ lbs.

*Hint: $\frac{1000\text{ kg}}{1} \times \frac{2\text{ lb.}}{1\text{ kg}} = ?$ lbs.

274. A baby weighing 8 *lbs.* would have a mass of about ____*____ kg.

*Hint: $\frac{8\text{ lb.}}{1} \times \frac{1\text{ kg}}{2\text{ lb.}} = ?$ kg

275. One kg of water takes up____________cm^3 of space. 5000 cm^3 of water weigh____________ g, or ____________ kg. (Ignore the 4°C. requirement which only slightly affects the relationship.)

276. 1000 cm^3 of water have a mass of ____________ kg, and weigh about ____________ lbs.

277. 6000 $cm^3$of water have a mass of ____________ kg. Since a quart is about the size of 1000 cm^3, then 6000 cm^3 are about ____________ qts. of water. The 6000 cm^3 weigh about ____________ lbs. (2 lbs. per 1 kg)

278. 8000 g of water weigh about ____________ lbs. 8000 g of water are about ____________ qts.

279. ____________ lbs. is roughly the mass of 1 kg?

____________ lb. is roughly the mass of 500 g?

____________ lb. is roughly the mass of 0.5 kg?

280. 500 cm^3 of water are about ____________ quarts, and weigh about ____________ pound.

281. If a man weighs 200 pounds, he has a mass of about ____________ kg.

282. When a person dives into a swimming pool his body pushes aside (displaces) the water that would be in his place if he had not entered the pool. The volume of water displaced will be equal to the volume of his ____________.

283. If a person displaces 50 quarts of water, his body must have a volume of ____________ quarts.

284. Some people neither sink to the bottom nor float on the surface, but remain submerged just under the surface. This happens when the total mass of their body is the same as the total mass of the ____________ they displace. Their mass is the same as the water displaced and they will neither ____________ to the bottom nor ____________ to the surface.

285. In such a case, the ____________ of the person's body is the same as the mass of the water displaced. If 100 000 cm^3 of water were displaced, the mass of the displaced water would be ____________ g, or ____________ kg, and the mass of the person would be ____________ kg.

286. A person who displaced 100 000 cm^3 of water having a mass of 100 kg would also have a volume of ____________ cm^3, and a mass of ____________ kg, if he neither sinks nor floats on the surface.

287. Most people come very close to just floating or just sinking in water. Therefore, if we know their ____________ in kg it is easy to estimate the amount of ____________ they take up in cubic centimeters.

288. A gram of water takes up 1 cm^3 of space and generally a gram of human tissue takes up ____________ of space.

289. 1000 grams of human tissue would take up ____________ cm^3 of space. Two kg of human tissue would take up ____________ cm^3 of space. A student with a mass of 50 kg would take up ____________ of space.

290. If a person weighs about 200 pounds, approximately how much space does he fill? Show your work.*

*Hint: $\frac{200 \text{ lbs.}}{1} \times \frac{1 \text{ kg}}{2 \text{ lb.}} \times \frac{1000 \text{ cm}^3}{1 \text{ kg}} = ?$

291. A kilogram of water (or human tissue) takes up about 1 quart of space. How many quarts of space are taken up by a person weighing 100 lb.? Show your work.

$\frac{100 \text{ lb.}}{1} \times \frac{? \text{ kg}}{? \text{ lb.}} \times \frac{? \text{ qt.}}{? \text{ kg}} = ?$

292. A person takes up about 1000 cm^3 or 1 qt. of space for each kg of mass. If a hundred people with an average weight of 100 lbs. each jumped into a swimming pool that was ready to overflow, how many gallons of water would run out? Show work.

$\frac{? \text{ persons}}{1} \times \frac{? \text{ lbs.}}{1 \text{ person}} \times \frac{? \text{ kg}}{? \text{ lb.}} \times \frac{? \text{ qt.}}{? \text{ kg}} \times \frac{1 \text{ gal.}}{4 \text{ qt.}} = ?$

293. If the swimming pool only held 1000 gallons of water, you can see that 100 people, at 100 lbs. each, would make this pool more than simply crowded! Now review: A gram is a metric unit designed to measure ____________. One gram is often defined as the mass of ____________ cm^3 of ____________ at ____________ temperature.

294. There is another very small mass unit which is sometimes used. This unit is the milligram. 1000 ____________ equal 1 gram. The relationship of milligrams to grams is a ratio of ____________ (as fraction and label).

295. The symbol for ____________ is mg. 1000 mg of water would take up ____________ cm^3 of space.

296. One mg is equal to one part of a gram of mass that is divided into ____________ equal parts.

297. 500 mg would be the same mass as ____________ g. 2000 mg is the same as ____________ g; a kilogram contains how many milligrams?

$$\frac{1\ \not{kg}}{1} \times \frac{1000\ g}{1\ \not{kg}} \times \frac{\quad\quad\quad}{?} = ?$$

298. Milligram units are so ____________ that it is very difficult to estimate mass measured in milligrams.

299. A nickel has a mass of approximately 5 g (but a penny unfortunately *does not* have a mass of 1 g). Two nickels would have a mass of ____________ g. 100 nickels would have a mass of ____________ g, or ____________ kg.

300. If an apple has the same approximate mass as 8 nickels, what is the mass of the apple? ____________.

301. A 10 pound bag of sugar has a mass of approximately ____________ kg, or ____________ g.

302. A 1 pound steak would be approximately ____________ kg, or ____________ g.

303. The standard metric unit for measuring volume is a unit called the liter (pronounced lē·ter). The English system units for the measurement of volume are units like the cubic foot, cubic yard, cup, pint, quart, gallon, etc. The basic unit for measuring volume in the metric system is the ____________.

304. The size of 1 liter is easy to grasp. A volume of one l____________ is the space occupied by a cube measuring 1 decimeter on each side. The accompanying illustration shows such a cube having a volume of ____________ ____________. Very minor corrections in the original measurement have changed the liter measurement to 1.000 028 cubic decimeters. This small error is unimportant in all but precise measurements.

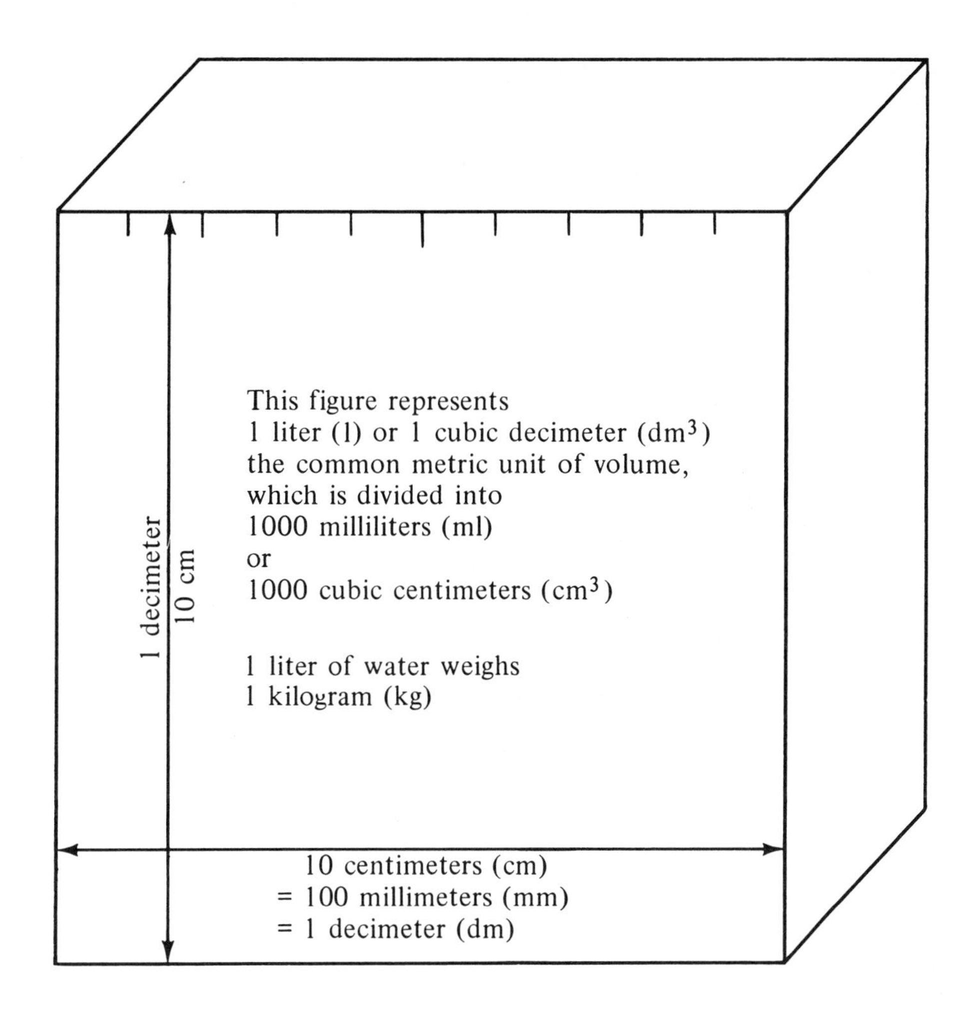

305. If a liter is a cube, 1 dm by 1 dm by 1 dm, it is also a cube that measures ____________ centimeters by ____________ cm by ____________ cm.

306. If a liter container is 10 cm high by 10 cm long by 10 cm wide, then it contains 10 cm X 10 cm X 10 cm, or a total of ____________ cubic centimeters.

307. If a liter contains ____________ cubic centimeters, you could place a row of ____________ cubic centimeter cubes along the front edge of a liter container. A series of ____________ rows would be needed to cover the bottom. Finally, ____________ layers containing 100 cubic centimeter cubes each would completely fill the liter container. (See the drawings on pages 79 and 117.)

308. A liter of water would be the same volume as ____________ cm^3 of water. A liter of water would, therefore, have a mass of 1 ____________.

309. A liter of milk would nearly fill a quart container. When only approximate relationships are necessary you may state that the ratio of a liter to a quart is $\frac{?\ \text{liter}}{1\ \text{quart}}$.

310. The relationship of a liter to a cubic centimeter is ———— $\frac{?\ \text{liter}}{?\ \text{cm}^3}$.

This figure represents 1 milliliter (ml)
or 1 cubic centimeter (cm^3)
1 ml of water weighs 1 gram (g)

311. A cube holds a liter of water having a mass of 1 kilogram. If the cube is divided into 1000 equal parts, each one will have a mass of ____________ gram. Each of the 1000 small cubes will have a volume of ____________ cm^3. (See drawings on pages 79 and 117.)

312. Like the meter unit and the gram unit, the liter unit can be subdivided into 1000 equal parts called ____________liters. A milliliter is considered to be 1 cubic centimeter.

313. Since a liter contains 1000 ml, and very nearly 1000 cm^3, then 1 ml is very nearly the same volume as ____________ cm^3.

314. A cube which has 1000 cm^3 units in a block has the same dimensions on each side. This large block represents 1000 cm^3 or 1000 ____________, or 1 ____________.(See drawings on pages 79 and 117.)

315. Gasoline is sold in many countries by the liter. A fuel tank in a foreign made car holds 40 liters. Since a liter is about ____________ quart, and since there are 4 quarts per gallon, this fuel tank holds about ____________ gallons.

316. A 10 gallon tank would hold about ____________ l, or ____________ ml of gasoline.

317. A 500 ml glass beaker would hold about 1 pint of liquid, or ____________ liter.

318. A can of soup is often 1/2 pint in volume. Since 500 cm^3 is approximately 1 pint, then the soup can has a volume of approximately ____________ cm^3.

319. The standard unit of volume in the metric system is the unit called a ____________. This unit can be considered the same as the volume of ____________ cm^3.

320. A cm^3 unit is the volume of a cube ____________ ____________ on each edge.

END OF UNIT 7

ANSWERS

Unit 7 DIMENSIONS OF MASS AND VOLUME Frames 244-320

244. ten (10); clue

245. pound

246. earth; force; pound

247. gravity; force; pounds

248. earth; force; weight

249. force; one pound

250. distance (amount); pound

251. 0; weight

252. force; gravity

253. gravity

254. gravity; force

255. Pikes Peak

256. weight; gravity

257. gram; mass

258. mass; distance (length)

259. one centimeter

260. contracts

261. 4°C.; 4°C.

262. 4°C. (39°F.); gram

263. 1 cm; water; 4°C. (39°F.)

264. cubic centimeter; cm^3

265. mass; 1 gram

266. lead; lead

267. water; air

268. 1 cm^3; less

269. 1000; 75 000

270. 1000; 1000 cm^3

271. 2 lbs. (drop the .2 for this work); 1 kg; 2 lb.

272. 2 lbs.; 100 lbs.

273. 2000 lbs. (1 ton)

274. 4 kg (2 lbs. for each kg)

275. 1000 cm^3; 5000 g; 5 kg

276. 1 kg; 2 lbs.

277. 6 kg; 6 qts.; 12 lbs.

278. 16 lbs.; 8 qts.

279. 2 lbs.; 1 lb.; 1 lb.

280. 1/2 or .5 qt.; 1 lb.

281. 100 kg

282. body

283. 50 quarts

284. water; sink; float

285. mass; 100 000 g; 100 kg; 100 kg

286. 100 000 cm^3; 100 kg

287. mass; space (volume)

288. one cm^3

289. 1000 cm^3; 2000 cm^3; 50 000 cm^3

290. 100 000 cm^3

291. $$\frac{\overset{50}{\cancel{100\text{ lb.}}}}{1} \times \frac{1\ \cancel{\text{kg}}}{\underset{1}{\cancel{2\text{ lb.}}}} \times \frac{1\text{ qt.}}{\cancel{1\text{ kg}}} = 50\text{ qts.}$$

292. $$\frac{100\text{ persons}}{1} \times \frac{\overset{50}{\cancel{100\text{ lb.}}}}{1\ \cancel{\text{person}}} \times \frac{1\ \cancel{\text{kg}}}{\underset{1}{\cancel{2\text{ lb.}}}} \times \frac{1\ \cancel{\text{qt.}}}{1\ \cancel{\text{kg}}}$$
$$\times \frac{1\text{ gal.}}{4\ \cancel{\text{qt.}}} = \frac{5000\text{ gal.}}{4} \text{ or } 1250\text{ gal. of water displaced.}$$

293. mass; 1 cm^3; pure water; 4°C.

294. milligrams; $\frac{1000\text{ milligrams}}{1\text{ gram}}$

295. milligram; 1 cm^3

296. 1000

297. 0.5 or 1/2 g; 2 g;

$$\frac{1\text{ kg}}{1} \times \frac{1000\text{ g}}{1\text{ kg}} \times \frac{1000\text{ mg}}{1\text{ g}} = 1\ 000\ 000\text{ mg}$$

298. small

299. 10 g; 500 g; 0.5 g or 1/2 kg

300. 40 g

301. 5 kg; 5000 g

302. 0.5 or 1/2 kg; 500 g

303. liter

304. liter; one liter

305. 10 cm; 10 cm; 10 cm

306. 1000

307. 1000 cm^3; 10 cm^3; 10; 10

308. 1000 cm^3; kilogram

309. 1

310. $\frac{1\text{ liter}}{1000\text{ cm}^3}$

311. one; one

312. milliliters

313. one

314. ml; liter

315. one; ten

316. 40 l; 40 000 ml

317. 1/2

318. 250 cm^3

319. liter; 1000 cm^3

320. one centimeter

Think Metrically

The comparative dimensions of the liter and the quart, and the kilogram and the pound are shown.

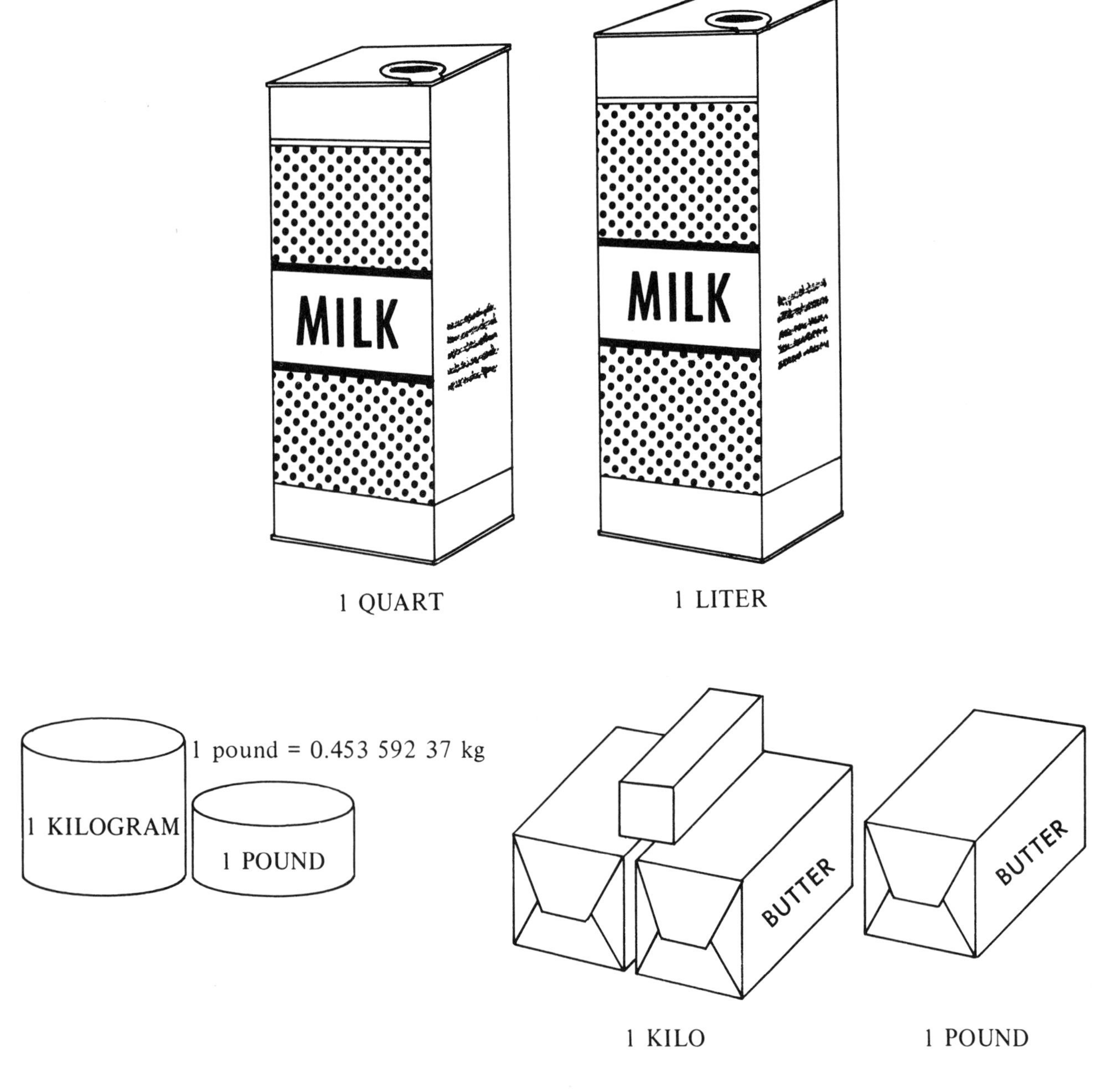

Unit 8

TESTING YOUR SKILLS (REVIEW)

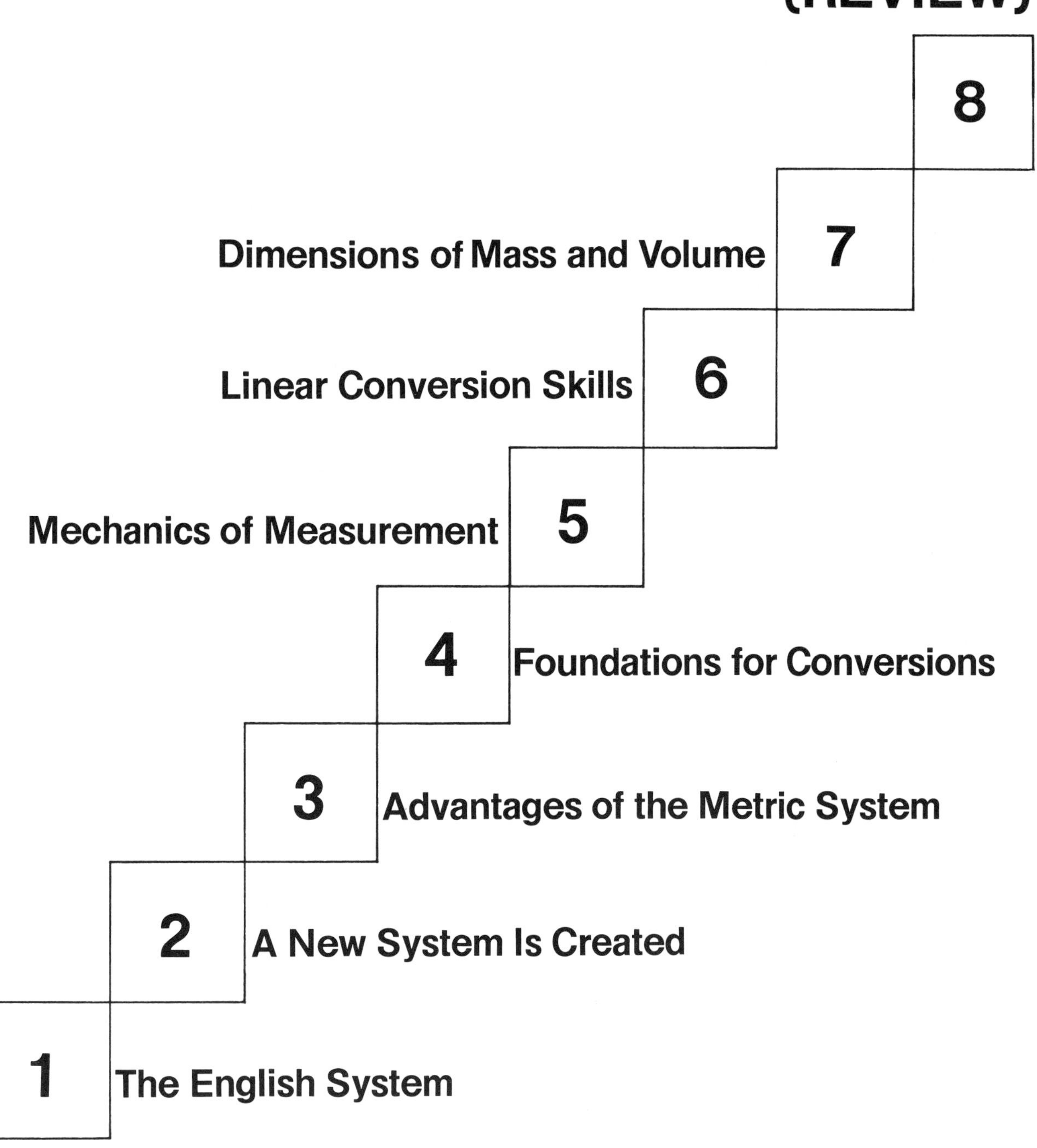

Introduction

Can you begin to "see" volume and mass in metric units? Try these:

1. Volume of 5 grams of water.
2. Mass of 1 liter of water.
3. 250 cm^3 of H_2O.
4. A 2 cm^3 injection.
5. 40 liters of gas bought in Canada.
6. Mass of a $.50 piece.
7. Cm^3 in a cup of milk.
8. Grams in a cup of milk.
9. Mass of a large bowling ball.
10. Volume of a ping-pong ball.
11. Mass of a tennis ball.

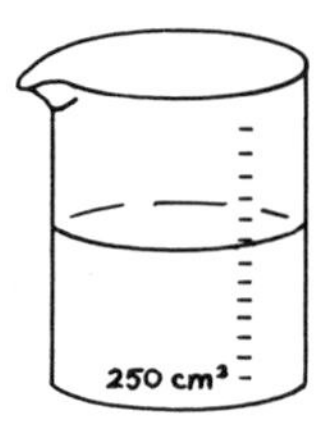

Turn to page 92 for the answers to Unit 8, Frames 321-353.

321. The English system is not a planned system of measurement. Two major weaknesses make it difficult to use. The ____________ of one unit to another is never a simple power of ____________, and none of the English unit names have ____________ to show their relative size.

322. The relationship of one unit of measurement to another unit of measurement can be expressed as a ____________ in which the value of the numerator is ____________ to the value of the denominator.

DIRECTIONS – For each of the following items through frame number 333, write a ratio expressing the relation of the first unit to the second.

323. decimeter to meter

324. centimeter to meter

325. millimeter to meter

326. meter to kilometer

327. meter to millimeter

328. centimeter to decimeter

329. millimeter to centimeter

330. kilogram to gram

331. milliliter to liter

332. cubic centimeter to milliliter

333. liter to cubic decimeter

DIRECTIONS – Select the best of the several possible answers.

334. The distance of one meter is approximately:

A. 40 inches
B. 2 1/2 feet
C. 5 feet
D. less than one yard

335. The distance of one millimeter is approximately:

A. the length of a common pencil
B. the thickness of a dollar bill
C. the thickness of a dime
D. the thickness of a fifty-cent piece

336. The distance of one centimeter is approximately:

A. the diameter of a fifty-cent piece
B. the length of a common pencil
C. a little more than the diameter of a pencil
D. the length of a dollar bill

337. The distance of a kilometer is approximately:

A. the width of a school desk
B. the length of a typical school room
C. a little over one-half mile
D. over one mile

338. The mass of one gram is approximately:

A. two pounds
B. one-half pound
C. the same as a quart of water
D. the same as a cubic centimeter of water

339. The mass of a kilogram is approximately:

A. two pounds
B. less than one pound
C. the same as one ton
D. the same as a fifty-cent piece

340. The liter is approximately the same volume as:

A. a gallon jug
B. a quart jar
C. a half pint bottle
D. the space inside a small thimble

341. The cubic centimeter is about the same volume as:

A. one liter
B. the space inside a common glass playing marble
C. a gallon jug
D. the space taken up by the head of one housefly

342. One student in secondary school might likely have a mass of:

A. 5 kilograms
B. 55 kilograms
C. 555 kilograms
D. 5555 kilograms

343. One college student might likely have a height of:

A. 1 meter
B. 150 centimeters
C. 3 meters
D. 500 centimeters

DIRECTIONS – Convert the first measurement to the second unit. Use the formal conversion method.

344. 15 meters to ____________ centimeters

345. 36 460 000 meters to ____________ kilometers

346. 589 746 meters to ____________ millimeters

347. 417 kilometers to ____________ centimeters

348. 55 kilograms to ____________ grams

349. 484 736 850 millimeters to ____________ kilometers

350. 1200 inches to ____________ millimeters (use 40 inches to m)

351. 365 days to ____________ seconds

352. Using the approximate relationship of 2 pounds to 1 kilogram, and 2000 pounds to 1 ton, change 60 000 000 grams to ____________ tons.

353. How many cubic centimeters of water would be required to equal the mass of a person weighing 160 lbs.? Use the formal method of conversion to find the answer.

END OF UNIT 8

ANSWERS

Unit 8 TESTING YOUR SKILLS Frames 321-353

321. relationships (ratio); ten; prefixes (clues)

322. ratio; equal

323. $\frac{\text{10 decimeters}}{\text{1 meter}}$

324. $\frac{\text{100 centimeters}}{\text{1 meter}}$

325. $\frac{\text{1000 millimeters}}{\text{1 meter}}$

326. $\frac{\text{1000 meters}}{\text{1 kilometer}}$

327. $\frac{\text{1 meter}}{\text{1000 millimeters}}$

328. $\frac{\text{10 centimeters}}{\text{1 decimeter}}$

329. $\frac{\text{10 millimeters}}{\text{1 centimeter}}$

330. $\frac{\text{1 kilogram}}{\text{1000 grams}}$

331. $\frac{\text{1000 milliliters}}{\text{1 liter}}$

332. $\frac{\text{1 cubic centimeter}}{\text{1 milliliter}}$

333. $\frac{\text{1 liter}}{\text{1 cubic decimeter}}$

334. A

335. C

336. C

337. C

338. D

339. A

340. B

341. B

342. B

343. B

344. $\frac{\overset{15}{\cancel{15\text{ m}}}}{1} \times \frac{100\text{ cm}}{\underset{1}{\cancel{1\text{ m}}}} = 1500\text{ cm}$

345. $\frac{\overset{36\,460}{\cancel{36\,460\,000\text{ m}}}}{1} \times \frac{1\text{ km}}{\underset{1}{\cancel{1000\text{ m}}}} = 36\,460\text{ km}$

346. $\frac{\overset{589\,746}{\cancel{589\,746\text{ m}}}}{1} \times \frac{1000\text{ mm}}{\underset{1}{\cancel{1\text{ m}}}} = 589\,746\,000\text{ mm}$

347. $\frac{\overset{417}{\cancel{417\text{ km}}}}{1} \times \frac{\overset{1000}{\cancel{1000\text{ m}}}}{\underset{1}{\cancel{1\text{ km}}}} \times \frac{100\text{ cm}}{\underset{1}{\cancel{1\text{ m}}}} =$

41 700 000 cm

348. $\frac{\overset{55}{\cancel{55\text{ kg}}}}{1} \times \frac{1000\text{ g}}{\underset{1}{\cancel{1\text{ kg}}}} = 55\,000\text{ g}$

349.

$$\frac{\overset{\overset{484.736\,850}{\cancel{484\,736.850}}}{\cancel{484\,736\,850\text{ mm}}}}{1} \times \frac{\overset{1}{\cancel{1\text{ m}}}}{\underset{1}{\cancel{1000\text{ mm}}}} \times \frac{1\text{ km}}{\underset{1}{\cancel{1000\text{ m}}}}$$

= 484.736 850 km

350.

$$\frac{\overset{30}{\cancel{1200\text{ inches}}}}{1} \times \frac{\overset{1}{\cancel{1\text{ m}}}}{\underset{1}{\cancel{40\text{ inches}}}} \times \frac{1000\text{ mm}}{\underset{1}{\cancel{1\text{ m}}}} =$$

30 000 mm

351.

$$\frac{\overset{365}{\cancel{365\text{ days}}}}{1} \times \frac{\overset{24}{\cancel{24\text{ hr.}}}}{\underset{1}{\cancel{1\text{ day}}}} \times \frac{\overset{60}{\cancel{60\text{ min.}}}}{\underset{1}{\cancel{1\text{ hr.}}}} \times \frac{60\text{ sec.}}{\underset{1}{\cancel{1\text{ min.}}}} =$$

31 536 000 sec.

352.

$$\frac{\overset{\overset{60}{\cancel{60\,000}}}{\cancel{60\,000\,000\text{ g}}}}{1} \times \frac{\overset{1}{\cancel{1\text{ kg}}}}{\underset{1}{\cancel{1000\text{ g}}}} \times \frac{\overset{1}{\cancel{2\text{ lb.}}}}{\underset{1}{\cancel{1\text{ kg}}}} \times \frac{1\text{ ton}}{\underset{\underset{1}{\cancel{1000}}}{\cancel{2000\text{ lb.}}}} = 60\text{ tons}$$

353.

$$\frac{\overset{80}{\cancel{160\text{ lbs.}}}}{1} \times \frac{\overset{1}{\cancel{1\text{ kg}}}}{\underset{1}{\cancel{2\text{ lbs.}}}} \times \frac{\overset{1000}{\cancel{1000\text{ g}}}}{\underset{1}{\cancel{1\text{ kg}}}} \times \frac{1\text{ cm}^3}{\underset{1}{\cancel{1\text{ g}}}} =$$

80 000 cm^3

APPENDICES

Appendix A

METRIC LABORATORY PRACTICE

These simple exercises are designed to improve your "feeling" for metric measurements.

DIRECTIONS

1. Make a calculated guess about the approximate answer to each question listed.
2. After you have guessed the approximate answer get the laboratory equipment you will need to measure and find the correct answer. You need not bother with fractions of cm^3, grams, etc., unless, of course, the correct answer is only part of a cm^3 or gram.
3. Complete both the guess and the exact measurement of each question before going to the next problem.
4. Guess and measure all problems in metric units.
5. Prepare your own answer sheet with two columns, the first column your guess, the second the correct answer.

Example:	*Guess*	*Correct*
1.	.5 mm	1.5 mm

   Try to come as close to the correct measurement as possible by gradually improving your ability to guess approximate amounts.
6. Check any problems with which you have difficulty with a classmate or your teacher.

PROBLEMS

Linear

Give all answers in metric units.

1. What is the width of your pencil or pen point?
2. How high is your desk or lab table top from the floor?
3. What is the width of your right thumb nail?
4. What is the width of the answer sheet you are using?
5. What is the length and width of the room you are in?

6. How tall are you?
7. What is the circumference of your waist?
8. How long is the longest hall in your school building?
9. How far is it from home plate to second base on the softball diamond you use?
10. How thick is a half-dollar?
11. How thick is ten sheets of paper?
12. How high is a tennis net from the ground?
13. What is the width of home plate on a baseball diamond?
14. What is the diameter of the opening in the nearest pencil sharpener?
15. What is the length of this sentence?

Mass

Give all answers in metric units.

1. What is the mass of your pen or pencil?
2. What is the mass of this complete manual?
3. How many kilograms of mass are you? (convert from English)
4. What mass has a blackboard eraser?
5. What mass has a new piece of chalk?
6. What mass has a half-dollar?
7. What mass have 250 ml of water?
8. What mass has a liter of water?
9. What is the mass of a 250 ml beaker?
10. What is the mass of a meter stick?
11. What is the mass of a petri dish?
12. What is the mass of a pound of feathers?
13. What is the mass of a ping-pong ball?
14. What is the mass of your pocket comb?
15. What is the mass of 10 drops of water?

Volume

Give all answers in metric units.

1. What is the volume of a half-dollar?
2. What is the volume of a kilogram of water?
3. What is the volume of a new piece of chalk?

4. Pour an unknown amount of water into a water glass. What is the volume of water in the container?
5. Fill a pail, bucket, or pneumatic trough with water. What is the volume of water in the container?
6. What is the volume of 10 drops of water?
7. Find a cork stopper. Find its volume.
8. Put a stopper in the top of an Erlenmeyer flask or in a gas bottle. What is the volume of air in the container?
9. What is the volume of a dime?
10. What is the volume of a pint of water?
11. What is the volume of a ten gallon hat?
12. Find a soda straw. What is the volume of air that will fit inside the straw?
13. Find the volume of a ping-pong ball.
14. What is the volume of 2 teaspoons full of NaCl?
15. Locate a dry sponge. Find the true volume of the sponge minus the holes.

Appendix B

EXPONENTS – SIMPLIFIED NUMERALS

EXPONENTIAL NOTATION
(Powers of 10)

In working with large numbers in either metric or English units you need to be familiar with powers of ten. This is called expressing numbers in *exponential notation.*

A measuring term known as a mole contains 602 300 000 000 000 000 000 000 molecules. Rather than say or write 602 sextillion, 300 quintillion, it would be far easier to simply say 6.023×10^{23}. Needless to say, multiplication or division would be a nightmare if such large numbers could not be reduced and handled in a simple manner.

Equally difficult small numbers can be handled with simplicity. The mass of an electron, for example, is about 0.000 000 000 000 000 000 000 000 911 grams! Adding or subtracting such small numbers would be laboriously dull. The same number, however, as 9.11×10^{-28} can be handled nicely.

Learning to express numbers by their exponents is quite simple. Problems are provided for your practice. Try them.

These guides should help you get started. Common powers of 10 can be expressed as:

10^0	=	1			
10^1	=	10	or	0.1	= 10^{-1}
10^2	=	100	or	0.01	= 10^{-2}
10^3	=	1000	or	0.001	= 10^{-3}
10^4	=	10 000	or	0.000 1	= 10^{-4}

PROBLEMS WITH EXPONENTS

Directions:

Solve each item. See page 102 for answers to these problems.

1. $10^2 \times 10^3 =$
2. $10^6 \times 10^1 =$
3. $\frac{10^4}{10^2} =$
4. $\frac{10^{12}}{10^{10}} =$

5. $(4.2 \times 10^2) \times (5.0 \times 10^4) =$

6. $(3.1 \times 10^5) \times (2.0 \times 10^3) =$

7. $\dfrac{8.0 \times 10^6}{4.0 \times 10^3} =$

8. $\dfrac{9.2 \times 10^{12}}{4.0 \times 10^3} =$

9. $10^5 \times 10^{-2} =$

10. $10^{12} \times 10^{-8} =$

11. $10^5 \times 10^{-10} =$

12. $10^3 \times 10^{-7} =$

13. $(4.3 \times 10^4) \times (2.0 \times 10^{-2}) =$

14. $(8.0 \times 10^{-6}) \times (1.0 \times 10^{-2}) =$

15. $\dfrac{1.8 \times 10^4}{2.0 \times 10^7} =$

16. $\dfrac{3.2 \times 10^{10}}{2.0 \times 10^{-4}} =$

17. $\dfrac{5.2 \times 10^{-8}}{2.0 \times 10^4} =$

18. $\dfrac{8.4 \times 10^{-8}}{4.0 \times 10^{-2}} =$

19. $\dfrac{9.8 \times 10^4}{0.2 \times 10^2} =$

20. $\dfrac{0.6 \times 10^3}{3.0 \times 10^1} =$

21. $4.5 \times 10^9 \div 5 \times 10^4 =$

22. $1.77 \times 10^8 \div 3 \times 10^{-3} =$

23. $1.45 \times 10^{-8} \div 5.0 \times 10^{-12} =$

24. $2.3 \times 10^4 + 4.6 \times 10^4 =$

25. $2.3 \times 10^4 + 5.3 \times 10^5 =$

26. $2.18 \times 10^{-2} - 1.09 \times 10^{-2} =$

27. $6.23 \times 10^{-9} - 8.5 \times 10^{-10} =$

28. $116 \times 10^5 = 1.16 \times 10^?$

29. $2356 \times 10^5 = 2.356 \times 10^?$

30. $448 \times 10^4 = 4.48 \times 10^?$

31. $0.029 \times 10^5 = 2.9 \times 10^?$

32. $225 \times 10^5 = 2.25 \times 10^?$

33. $0.054 \times 10^{-9} = 5.4 \times 10^?$

ANSWERS TO EXPONENT PROBLEMS

(pages 100–101)

1. 10^5
2. 10^7
3. 10^2
4. 10^2
5. 2.1×10^7
6. 6.2×10^8
7. 2.0×10^3
8. 2.3×10^9
9. 10^3
10. 10^4
11. 10^{-5}
12. 10^{-4}
13. 8.6×10^2
14. 8.0×10^{-8}
15. 0.9×10^{-3} or *9.0×10^{-4}
16. 1.6×10^{14}
17. 2.6×10^{-12}
18. 2.1×10^{-6}
19. 49.0×10^2 or *4.9×10^3
20. 0.2×10^2 or *2.0×10^1
21. 0.9×10^5 or *9.0×10^4
22. 0.59×10^{11} or *5.9×10^{10}
23. 0.29×10^4 or *2.9×10^3
24. 6.9×10^4
25. 55.3×10^4 or *5.53×10^5
26. 1.09×10^{-2}
27. 5.38×10^{-9}
28. 1.16×10^7
29. 2.356×10^8
30. 4.48×10^6
31. 2.9×10^3
32. 2.25×10^7
33. 5.4×10^{-11}

---

*this form preferred

Appendix C

PROFICIENCY TEST

The following 55 questions represent a cross section of all units of *The Metric System* as follows:

Questions	*Topic*
1-4	Abbreviations
5-29	Metric to Metric Conversions
30-40	Exponents
41-44	Conversion Problems
45-55	"Thinking Metric"

Directions:

1. *Select a test.* Turn to page 109 and select any one of the fourteen (forms A-0) tests. Each test has twenty-five questions. You do *not* attempt all questions, only those on the particular test you select.

2. *Take the test.* All questions are multiple choice. Answer only the twenty-five questions on your test by placing an (X) over the best choice.

3. *Score your results.* Turn to the answers on page 112. A score of twenty-one or more correct indicates good proficiency. You are probably in good shape to understand and to use metric measurement.

4. *Scores less than twenty-one:*

   a) You may take the test as often as you like! Before you do, however, determine if your incorrect answers fall into a single category such as exponents (q. 30-40) or conversion problems (q. 41-44).

      Use *The Metric System* as a review. Re-do any of the eight units which you think would help. Doing some of the problems in Appendices A and B would probably help most.

   b) Now select a new test. Continue this procedure until you reach a proficiency satisfactory to you or to the teacher.

TEST

Multiple Choice: Answer only the *marked* items on the answer sheet. Mark the number of the best answer on the answer sheet provided. Do *not* write on this paper. Do *not* attempt all questions.

1. The symbol for millimeter is:
   1) m 2) dm 3) cm 4) mm
2. The symbol for centimeter is:
   1) m 2) cm 3) dm 4) mm
3. The symbol for decimeter is:
   1) m 2) km 3) cm 4) dm
4. The symbol for kilometer is:
   1) m 2) km 3) cm 4) dm
5. $\frac{1 \text{ km}}{1000 \text{ m}} =$
   1) $\frac{1000 \text{ cm}}{1 \text{ m}}$ 2) $\frac{1000 \text{ m}}{1 \text{ km}}$ 3) $\frac{1000 \text{ dm}}{1 \text{ cm}}$ 4) $\frac{1000 \text{ dm}}{1 \text{ km}}$
6. $\frac{100 \text{ cm}}{10 \text{ dm}} =$
   1) $\frac{1 \text{ m}}{10 \text{ dm}}$ 2) $\frac{1 \text{ dm}}{10 \text{ m}}$ 3) $\frac{1000 \text{ mm}}{10 \text{ cm}}$ 4) $\frac{100 \text{ cm}}{10 \text{ m}}$
7. $\frac{1 \text{ dm}}{10 \text{ cm}} =$
   1) $\frac{10 \text{ m}}{1 \text{ km}}$ 2) $\frac{10 \text{ cm}}{1 \text{ km}}$ 3) $\frac{10 \text{ cm}}{1 \text{ dm}}$ 4) $\frac{10 \text{ mm}}{1 \text{ dm}}$
8. $\frac{10 \text{ cm}}{100 \text{ cm}} =$
   1) $\frac{10 \text{ mm}}{100 \text{ mm}}$ 2) $\frac{1 \text{ mm}}{100 \text{ mm}}$ 3) .1 mm 4) 1 mm
9. This line ____ is approximately:
   1) 1 mm 2) 1 cm 3) 1 dm 4) 1 m
10. 10 dm equals
    1) 10 mm 2) 1 cm 3) 1 m 4) 10 m
11. $\frac{100 \text{ cm}}{1 \text{ m}} =$
    1) .01 2) .1 3) 1 4) 100
12. 100 cm equals
    1) 10 mm 2) 1000 mm 3) 100 mm 4) 10 000 mm
13. $\frac{10 \text{ dm}}{100 \text{ cm}} =$
    1) .001 2) .01 3) .1 4) 1

14. 1 mm equals
    1) .1 m 2) .01 m 3) .001 m 4) .0001 m

15. 1 km equals
    1) 10 m 2) 1 000 m 3) 10 000 m 4) 100 000 m

16. $\frac{1 \text{ m}}{10 \text{ dm}} =$
    1) 1 2) 10 3) $\frac{1}{10}$ 4) $\frac{1}{100}$

17. 100 km equals
    1) 10 m 2) 1 000 m 3) 10 000 m 4) 100 000 m

18. $\frac{10 \text{ dm}}{1 \text{ m}} =$
    1) 10 2) 100 3) $\frac{1}{10}$ 4) 1

19. $\frac{1\ 000\ 000 \text{ mm}}{1000} =$
    1) 1 cm 2) 1 dm 3) 1 m 4) 1 km

20. The relationship of a liter to cubic centimeters is:
    1) $\frac{1 \text{ L}}{1000 \text{ cm}^3}$ 2) $\frac{1 \text{ L}}{100 \text{ cm}^3}$ 3) $\frac{1 \text{ L}}{10 \text{ cm}^3}$ 4) $\frac{1 \text{ L}}{1 \text{ cm}^3}$

21. 4.32 liters of water equals:
    1) 43.2 cm^3 2) 432 cm^3 3) 4 320 cm^3 4) 43 200 cm^3

22. 87.325 liters of water weighs:
    1) 8.732 5 kg 2) 87.325 kg 3) 873.25 kg 4) 8 732.5 kg

23. A cm^3 unit is the volume of a cube ____________ ____________ on each side:
    1) 1 ml 2) 1 mm 3) 100 mm 4) 10 mm

24. A box measuring 15 mm by 10 mm by 10 cm would hold ____________ of water:
    1) 15 cm^3 2) .15 cm^3 3) 150 000 cm^3 4) 1 500 cm^3

25. A gram of water takes up ____________ of space:
    1) 1 cm^3 2) 10 cm^3 3) 100 cm^3 4) 1 000 cm^3

26. A kilogram of water takes up ____________ of space:
    1) 1 cm^3 2) 10 cm^3 3) 100 cm^3 4) 1 000 cm^3

27. A cube with the dimensions of 1 dm by 1 dm by 1 dm would have the volume of:
    1) 1 dm 2) 1 ml 3) 1 L 4) 1 kg

28. .001 L equals
    1) 1 ml 2) 1 cl 3) 1 dl 4) 1 L

29. 1 L of water equals
    1) 1 cm^3 2) 10 cm^3 3) 100 cm^3 4) 1000 cm^3

30. An irregular shaped rock placed in a container of water raises the water level 4.5 liters. The volume of the rock is:

1) 4.5×10 ml 2) 4.5×10^2 ml 3) 4.5×10^3 ml 4) 4.5×10^4 ml

31. 20 mg equals

1) 2.0×10^5 kg 2) 2.0×10^3 kg 3) 2.0×10^{-3} kg 4) 2.0×10^{-5} kg

32. $\frac{1.725 \times 10^3 \text{ cm}}{.5 \times 10^2 \text{ cm}} =$

1) 34.5 2) 3.45 3) 345 4) .345

33. $\frac{5.2 \times 10^{-8} \text{ m}}{2.0 \times 10^4 \text{ m}} =$

1) 2.6×10^{-4} 2) $3.815\,4 \times 10^{11}$ 3) 2.6×10^{-12} 4) 381.54×10^4

34. $\frac{9.8 \times 10^4 \text{ cm}}{0.2 \times 10^2 \text{ mm}} =$

1) 4.9×10^3 2) 4.9×10^4 3) 4.9×10^2 4) 4.9×10

35. To write this number (1.65×10^4 cm) *without* using exponents you would need to add:

1) 4 zeros before the digit "1"
2) 3 zeros before the digit "1"
3) 4 zeros after the digit "5"
4) 2 zeros after the digit "5"

36. 1.09×10^{-2} cm =

1) .001 09 cm 2) .0109 cm 3) 109 cm 4) 10 900 cm

37. 6.0×10^3 m =

1) 60×10^2 km 2) 6 km 3) .6 km 4) 60 km

38. 2.85×10^2 dm =

1) 285 cm 2) 285 dm 3) 285 m 4) 285 km

39. 4.13×10^{-4} km =

1) 4.13 mm 2) 4.13 cm 3) 4.13 dm 4) 4.13 m

40. 1.09×10^{-2} cm =

1) 1.09 m 2) 1.09×10^2 m 3) 1.09×10^{-4} m 4) 1.09 km

41. The conversion procedure for 1 kg to 1 000 000 mg =

1) $\frac{1 \text{ kg}}{1} \times \frac{1000 \text{ g}}{1 \text{ kg}} \times \frac{1000 \text{ mg}}{1 \text{ g}}$ 2) $\frac{1}{1 \text{ kg}} \times \frac{1 \text{ kg}}{1000 \text{ g}} \times \frac{1 \text{ g}}{1000 \text{ mg}}$

3) $\frac{1 \text{ kg}}{1} \times \frac{1 \text{ kg}}{1000 \text{ g}} \times \frac{1000 \text{ mg}}{1 \text{ g}}$ 4) $\frac{1}{1 \text{ kg}} \times \frac{1000 \text{ kg}}{1 \text{ g}} \times \frac{1000 \text{ g}}{1 \text{ mg}}$

42. The conversion procedure for 1000 g to 1 kg =

1) $\frac{1000 \text{ g}}{1} \times \frac{1 \text{ kg}}{1\,000\,000 \text{ mg}} \times \frac{1000 \text{ mg}}{1 \text{ g}}$ 2) $\frac{1}{1000 \text{ g}} \times \frac{1 \text{ kg}}{1\,000\,000 \text{ mg}} \times \frac{1000 \text{ mg}}{1 \text{ g}}$

3) $\frac{1 \text{ kg}}{100 \text{ g}} \times \frac{1000 \text{ g}}{10 \text{ kg}} \times \frac{1 \text{ kg}}{1}$ 4) $\frac{10 \text{ g}}{1 \text{ kg}} \times \frac{1000 \text{ mg}}{10 \text{ g}} \times \frac{1}{1000 \text{ kg}}$

43. The conversion procedure for 0.13×10^2 mm to 1.3 cm =

1) $\frac{13 \text{ mm}}{1} \times \frac{10 \text{ cm}}{1 \text{ mm}} \times \frac{1 \text{ m}}{1000 \text{ mm}}$
2) $\frac{100 \text{ cm}}{1 \text{ m}} \times \frac{13 \text{ mm}}{1} \times \frac{1 \text{ m}}{1000 \text{ mm}}$
3) $\frac{10 \text{ cm}}{1 \text{ m}} \times \frac{1000 \text{ mm}}{1 \text{ m}} \times \frac{13 \text{ mm}}{1}$
4) $\frac{13 \text{ mm}}{1} \times \frac{1 \text{ m}}{10\,000 \text{ mm}} \times \frac{100 \text{ cm}}{1 \text{ m}}$

44. The conversion procedure for 10 dm to 1000 mm =

1) $\frac{100 \text{ cm}}{1000 \text{ mm}} \times \frac{100 \text{ cm}}{10 \text{ dm}} \times \frac{10 \text{ dm}}{1}$
2) $\frac{1000 \text{ mm}}{100 \text{ cm}} \times \frac{100 \text{ cm}}{10 \text{ dm}} \times \frac{10 \text{ dm}}{1}$
3) $\frac{1}{10 \text{ dm}} \times \frac{10 \text{ cm}}{100 \text{ mm}} \times \frac{1 \text{ m}}{100 \text{ cm}}$
4) $\frac{1000 \text{ mm}}{10 \text{ dm}} \times \frac{10 \text{ dm}}{1 \text{ m}} \times \frac{1 \text{ m}}{1000 \text{ mm}}$

45. The distance of one centimeter is approximately:

1) the diameter of a fifty-cent piece
2) the length of a common pencil
3) a little more than the diameter of a pencil
4) the length of a dollar bill

46. The distance of one millimeter is approximately:

1) the length of a common pencil
2) the thickness of a dollar bill
3) the thickness of a dime
4) the thickness of a fifty-cent piece

47. The distance of a kilometer is approximately:

1) the width of a school desk
2) the length of a typical school room
3) a little over one-half mile
4) a little over one mile

48. One college girl might likely have a height of:

1) 1 meter
2) 150 centimeters
3) 3 meters
4) 500 centimeters

49. The cubic centimeter is about the same volume as:

1) one liter
2) the space inside a common glass playing marble
3) a gallon jug
4) the space taken up by the head of one housefly

50. A kilogram of steak would comfortably feed:

1) 1 adult 2) 4 adults 3) 40 adults 4) 4000 adults

51. With a classroom of 20 first graders how many liters of punch would you order for the school picnic?

1) 1000 2) 100 3) 10 4) 1

52. A good field goal kicker in football is most accurate from:

    1) 18 m 2) 6 ml 3) 350 m 4) 2500 km

53. A cube 1 dm on a side would pass through a round hole with a diameter of:

    1) 8.75 cm 2) 37.002 mm 3) 30 cm 4) 0.07 m

54. If a container with the dimensions of 6.0×10^4 dm by 13.93×10^5 mm by 15 m were full of water it would most properly be called a:

    1) pitcher 2) lake 3) tea cup 4) swimming pool

55. A person weighing 100 kg diving into a swimming pool would displace water in the amount of:

    1) 10 m 2) 10 000 ml 3) 100 cm 4) 100 000 cm^3

Appendix D

QUESTION SHEETS

FORM A		FORM B		FORM C		FORM D		FORM E	
1.	1 2 3 4	3.	1 2 3 4	1.	1 2 3 4	2.	1 2 3 4	1.	1 2 3 4
2.	1 2 3 4	4.	1 2 3 4	2.	1 2 3 4	4.	1 2 3 4	4.	1 2 3 4
6.	1 2 3 4	5.	1 2 3 4	5.	1 2 3 4	5.	1 2 3 4	5.	1 2 3 4
9.	1 2 3 4	7.	1 2 3 4	8.	1 2 3 4	7.	1 2 3 4	7.	1 2 3 4
10.	1 2 3 4	11.	1 2 3 4	12.	1 2 3 4	8.	1 2 3 4	8.	1 2 3 4
13.	1 2 3 4	12.	1 2 3 4	14.	1 2 3 4	11.	1 2 3 4	10.	1 2 3 4
15.	1 2 3 4	16.	1 2 3 4	17.	1 2 3 4	12.	1 2 3 4	11.	1 2 3 4
16.	1 2 3 4	19.	1 2 3 4	19.	1 2 3 4	14.	1 2 3 4	12.	1 2 3 4
18.	1 2 3 4	21.	1 2 3 4	21.	1 2 3 4	17.	1 2 3 4	13.	1 2 3 4
20.	1 2 3 4	25	1 2 3 4	23.	1 2 3 4	19.	1 2 3 4	15.	1 2 3 4
22.	1 2 3 4	26.	1 2 3 4	25.	1 2 3 4	21.	1 2 3 4	24.	1 2 3 4
24.	1 2 3 4	27.	1 2 3 4	27.	1 2 3 4	23.	1 2 3 4	25.	1 2 3 4
28.	1 2 3 4	29.	1 2 3 4	29.	1 2 3 4	29.	1 2 3 4	29.	1 2 3 4
31.	1 2 3 4	30.	1 2 3 4	32.	1 2 3 4	32.	1 2 3 4	30.	1 2 3 4
33.	1 2 3 4	31.	1 2 3 4	35.	1 2 3 4	33.	1 2 3 4	33.	1 2 3 4
34.	1 2 3 4	35.	1 2 3 4	36.	1 2 3 4	36.	1 2 3 4	35.	1 2 3 4
37.	1 2 3 4	37.	1 2 3 4	38.	1 2 3 4	39.	1 2 3 4	37.	1 2 3 4
40.	1 2 3 4	38.	1 2 3 4	39.	1 2 3 4	40.	1 2 3 4	38.	1 2 3 4
41.	1 2 3 4	41.	1 2 3 4	42.	1 2 3 4	43.	1 2 3 4	43.	1 2 3 4
44.	1 2 3 4	43.	1 2 3 4	43.	1 2 3 4	44.	1 2 3 4	44.	1 2 3 4
47.	1 2 3 4	45.	1 2 3 4	45.	1 2 3 4	45.	1 2 3 4	45.	1 2 3 4
48.	1 2 3 4	46.	1 2 3 4	46.	1 2 3 4	46.	1 2 3 4	46.	1 2 3 4
50.	1 2 3 4	50.	1 2 3 4	49.	1 2 3 4	49.	1 2 3 4	48.	1 2 3 4
53.	1 2 3 4	51.	1 2 3 4	51.	1 2 3 4	54.	1 2 3 4	50.	1 2 3 4
55.	1 2 3 4	53.	1 2 3 4	52.	1 2 3 4	55.	1 2 3 4	51.	1 2 3 4

FORM F	FORM G	FORM H	FORM J	FORM K
1. 1 2 3 4	2. 1 2 3 4	2. 1 2 3 4	3. 1 2 3 4	1. 1 2 3 4
2. 1 2 3 4	4. 1 2 3 4	4. 1 2 3 4	4. 1 2 3 4	4. 1 2 3 4
7. 1 2 3 4	5. 1 2 3 4	7. 1 2 3 4	5. 1 2 3 4	5. 1 2 3 4
9. 1 2 3 4	8. 1 2 3 4	8. 1 2 3 4	11. 1 2 3 4	9. 1 2 3 4
10. 1 2 3 4	10. 1 2 3 4	11. 1 2 3 4	13. 1 2 3 4	10. 1 2 3 4
14. 1 2 3 4	12. 1 2 3 4	12. 1 2 3 4	16. 1 2 3 4	12. 1 2 3 4
15. 1 2 3 4	16. 1 2 3 4	16. 1 2 3 4	17. 1 2 3 4	13. 1 2 3 4
18. 1 2 3 4	17. 1 2 3 4	18. 1 2 3 4	19. 1 2 3 4	15. 1 2 3 4
20. 1 2 3 4	18. 1 2 3 4	20. 1 2 3 4	20. 1 2 3 4	16. 1 2 3 4
23. 1 2 3 4	21. 1 2 3 4	21. 1 2 3 4	24. 1 2 3 4	19. 1 2 3 4
24. 1 2 3 4	22. 1 2 3 4	24. 1 2 3 4	25. 1 2 3 4	21. 1 2 3 4
26. 1 2 3 4	26. 1 2 3 4	25. 1 2 3 4	28. 1 2 3 4	25. 1 2 3 4
27. 1 2 3 4	27. 1 2 3 4	33. 1 2 3 4	29. 1 2 3 4	26. 1 2 3 4
30. 1 2 3 4	34. 1 2 3 4	34. 1 2 3 4	32. 1 2 3 4	30. 1 2 3 4
31. 1 2 3 4	35. 1 2 3 4	36. 1 2 3 4	34. 1 2 3 4	33. 1 2 3 4
34. 1 2 3 4	36. 1 2 3 4	37. 1 2 3 4	36. 1 2 3 4	35. 1 2 3 4
37. 1 2 3 4	37. 1 2 3 4	38. 1 2 3 4	38. 1 2 3 4	37. 1 2 3 4
40. 1 2 3 4	38. 1 2 3 4	41. 1 2 3 4	40. 1 2 3 4	40. 1 2 3 4
41. 1 2 3 4	42. 1 2 3 4	43. 1 2 3 4	41. 1 2 3 4	41. 1 2 3 4
43. 1 2 3 4	44. 1 2 3 4	45. 1 2 3 4	44. 1 2 3 4	44. 1 2 3 4
46. 1 2 3 4	46. 1 2 3 4	46. 1 2 3 4	47. 1 2 3 4	48. 1 2 3 4
47. 1 2 3 4	48. 1 2 3 4	50. 1 2 3 4	48. 1 2 3 4	50. 1 2 3 4
49. 1 2 3 4	50. 1 2 3 4	53. 1 2 3 4	52. 1 2 3 4	52. 1 2 3 4
50. 1 2 3 4	51. 1 2 3 4	54. 1 2 3 4	54. 1 2 3 4	54. 1 2 3 4
54. 1 2 3 4	54. 1 2 3 4	55. 1 2 3 4	55. 1 2 3 4	55. 1 2 3 4

FORM L	FORM M	FORM N	FORM O
1. 1 2 3 4	3. 1 2 3 4	2. 1 2 3 4	2. 1 2 3 4
4. 1 2 3 4	4. 1 2 3 4	3. 1 2 3 4	3. 1 2 3 4
6. 1 2 3 4	8. 1 2 3 4	9. 1 2 3 4	5. 1 2 3 4
9. 1 2 3 4	11. 1 2 3 4	12. 1 2 3 4	6. 1 2 3 4
12. 1 2 3 4	12. 1 2 3 4	14. 1 2 3 4	10. 1 2 3 4
15. 1 2 3 4	14. 1 2 3 4	15. 1 2 3 4	11. 1 2 3 4
18. 1 2 3 4	19. 1 2 3 4	16. 1 2 3 4	13. 1 2 3 4
20. 1 2 3 4	21. 1 2 3 4	18. 1 2 3 4	14. 1 2 3 4
22. 1 2 3 4	23. 1 2 3 4	20. 1 2 3 4	19. 1 2 3 4
24. 1 2 3 4	26. 1 2 3 4	21. 1 2 3 4	21. 1 2 3 4
26. 1 2 3 4	27. 1 2 3 4	24. 1 2 3 4	22. 1 2 3 4
28. 1 2 3 4	29. 1 2 3 4	28. 1 2 3 4	23. 1 2 3 4
29. 1 2 3 4	31. 1 2 3 4	29. 1 2 3 4	28. 1 2 3 4
33. 1 2 3 4	33. 1 2 3 4	32. 1 2 3 4	30. 1 2 3 4
34. 1 2 3 4	36. 1 2 3 4	33. 1 2 3 4	31. 1 2 3 4
36. 1 2 3 4	39. 1 2 3 4	36. 1 2 3 4	32. 1 2 3 4
37. 1 2 3 4	40. 1 2 3 4	37. 1 2 3 4	35. 1 2 3 4
38. 1 2 3 4	41. 1 2 3 4	40. 1 2 3 4	36. 1 2 3 4
41. 1 2 3 4	42. 1 2 3 4	42. 1 2 3 4	41. 1 2 3 4
42. 1 2 3 4	45. 1 2 3 4	44. 1 2 3 4	42. 1 2 3 4
45. 1 2 3 4	50. 1 2 3 4	47. 1 2 3 4	46. 1 2 3 4
47. 1 2 3 4	51. 1 2 3 4	48. 1 2 3 4	47. 1 2 3 4
49. 1 2 3 4	53. 1 2 3 4	51. 1 2 3 4	50. 1 2 3 4
50. 1 2 3 4	54. 1 2 3 4	53. 1 2 3 4	52. 1 2 3 4
52. 1 2 3 4	55. 1 2 3 4	54. 1 2 3 4	53. 1 2 3 4

Appendix E

ANSWER SHEETS

FORM A	FORM B	FORM C	FORM D	FORM E
1. 1 2 3 ④	3. 1 2 3 ④	1. 1 2 3 ④	2. 1 ② 3 4	1. 1 2 3 ④
2. 1 ② 3 4	4. 1 ② 3 4	2. 1 ② 3 4	4. 1 ② 3 4	4. 1 ② 3 4
6. ① 2 3 4	5. 1 ② 3 4	5. 1 ② 3 4	5. 1 ② 3 4	5. 1 ② 3 4
9. 1 ② 3 4	7. 1 2 ③ 4	8. ① 2 3 4	7. 1 2 ③ 4	7. 1 2 ③ 4
10. 1 2 ③ 4	11. 1 2 ③ 4	12. 1 ② 3 4	8. ① 2 3 4	8. ① 2 3 4
13. 1 2 3 ④	12. 1 ② 3 4	14. 1 2 ③ 4	11. 1 2 ③ 4	10. 1 2 ③ 4
15. 1 ② 3 4	16. ① 2 3 4	17. 1 2 3 ④	12. 1 ② 3 4	11. 1 2 ③ 4
16. ① 2 3 4	19. 1 2 ③ 4	19. 1 2 ③ 4	14. 1 2 ③ 4	12. 1 ② 3 4
18. 1 2 3 ④	21. 1 2 ③ 4	21. 1 2 ③ 4	17. 1 2 3 ④	13. 1 2 3 ④
20. ① 2 3 4	25 ① 2 3 4	23. 1 2 3 ④	19. 1 2 ③ 4	15. 1 ② 3 4
22. 1 ② 3 4	26. 1 2 3 ④	25. ① 2 3 4	21. 1 2 ③ 4	24. ① 2 3 4
24. ① 2 3 4	27. 1 2 ③ 4	27. 1 2 ③ 4	23. 1 2 3 ④	25. ① 2 3 4
28. ① 2 3 4	29. 1 2 3 ④	29. 1 2 3 ④	29. 1 2 3 ④	29. 1 2 3 ④
31. 1 2 3 ④	30. 1 2 ③ 4	32. ① 2 3 4	32. ① 2 3 4	30. 1 2 ③ 4
33. 1 2 ③ 4	31. 1 2 3 ④	35. 1 2 3 ④	33. 1 2 ③ 4	33. 1 2 ③ 4
34. 1 ② 3 4	35. 1 2 3 ④	36. 1 ② 3 4	36. 1 ② 3 4	35. 1 2 3 ④
37. 1 ② 3 4	37. 1 ② 3 4	38. 1 ② 3 4	39. 1 2 ③ 4	37. 1 ② 3 4
40. 1 2 ③ 4	38. 1 ② 3 4	39. 1 2 ③ 4	40. 1 2 ③ 4	38. 1 ② 3 4
41. ① 2 3 4	41. ① 2 3 4	42. ① 2 3 4	43. 1 ② 3 4	43. 1 ② 3 4
44. 1 ② 3 4	43. 1 ② 3 4	43. 1 ② 3 4	44. 1 ② 3 4	44. 1 ② 3 4
47. 1 2 ③ 4	45. 1 2 ③ 4	45. 1 2 ③ 4	45. 1 2 ③ 4	45. 1 2 ③ 4
48. 1 ② 3 4	46. 1 2 ③ 4	46. 1 2 ③ 4	46. 1 2 ③ 4	46. 1 2 ③ 4
50. 1 ② 3 4	50. 1 ② 3 4	49. 1 ② 3 4	49. 1 ② 3 4	48. 1 ② 3 4
53. 1 2 ③ 4	51. 1 2 ③ 4	51. 1 2 ③ 4	54. 1 ② 3 4	50. 1 ② 3 4
55. 1 2 3 ④	53. 1 2 ③ 4	52. ① 2 3 4	55. 1 2 3 ④	51. 1 2 ③ 4

FORM F	FORM G	FORM H	FORM J	FORM K
1. 1 2 3 ④	2. 1 ② 3 4	2. 1 ② 3 4	3. 1 2 3 ④	1. 1 2 3 ④
2. 1 ② 3 4	4. 1 ② 3 4	4. 1 ② 3 4	4. 1 ② 3 4	4. 1 ② 3 4
7. 1 2 ③ 4	5. 1 ② 3 4	7. 1 2 ③ 4	5. 1 ② 3 4	5. 1 ② 3 4
9. 1 ② 3 4	8. ① 2 3 4	8. ① 2 3 4	11. 1 2 ③ 4	9. 1 ② 3 4
10. 1 2 ③ 4	10. 1 2 ③ 4	11. 1 2 ③ 4	13. 1 2 3 ④	10. 1 2 ③ 4
14. 1 2 ③ 4	12. 1 ② 3 4	12. 1 ② 3 4	16. ① 2 3 4	12. 1 ② 3 4
15. 1 ② 3 4	16. ① 2 3 4	16. ① 2 3 4	17. 1 2 3 ④	13. 1 2 3 ④
18. 1 2 3 ④	17. 1 2 3 ④	18. 1 2 3 ④	19. 1 2 ③ 4	15. 1 ② 3 4
20. ① 2 3 4	18. 1 2 3 ④	20. ① 2 3 4	20. ① 2 3 4	16. ① 2 3 4
23. 1 2 3 ④	21. 1 2 ③ 4	21. 1 2 ③ 4	24. ① 2 3 4	19. 1 2 ③ 4
24. ① 2 3 4	22. 1 ② 3 4	24. ① 2 3 4	25. ① 2 3 4	21. 1 2 ③ 4
26. 1 2 3 ④	26. 1 2 3 ④	25. ① 2 3 4	28. ① 2 3 4	25. ① 2 3 4
27. 1 2 ③ 4	27. 1 2 ③ 4	33. 1 2 ③ 4	29. 1 2 3 ④	26. 1 2 3 ④
30. 1 2 ③ 4	34. 1 ② 3 4	34. 1 ② 3 4	32. ① 2 3 4	30. 1 2 ③ 4
31. 1 2 3 ④	35. 1 2 3 ④	36. 1 ② 3 4	34. 1 ② 3 4	33. 1 2 ③ 4
34. 1 ② 3 4	36. 1 ② 3 4	37. 1 ② 3 4	36. 1 ② 3 4	35. 1 2 3 ④
37. 1 ② 3 4	37. 1 ② 3 4	38. 1 ② 3 4	38. 1 ② 3 4	37. 1 ② 3 4
40. 1 2 ③ 4	38. 1 ② 3 4	41. ① 2 3 4	40. 1 2 ③ 4	40. 1 2 ③ 4
41. ① 2 3 4	42. ① 2 3 4	43. 1 ② 3 4	41. ① 2 3 4	41. ① 2 3 4
43. 1 ② 3 4	44. 1 ② 3 4	45. 1 2 ③ 4	44. 1 ② 3 4	44. 1 ② 3 4
46. 1 2 ③ 4	46. 1 2 ③ 4	46. 1 2 ③ 4	47. 1 2 ③ 4	48. 1 ② 3 4
47. 1 2 ③ 4	48. 1 ② 3 4	50. 1 ② 3 4	48. 1 ② 3 4	50. 1 ② 3 4
49. 1 ② 3 4	50. 1 ② 3 4	53. 1 2 ③ 4	52. ① 2 3 4	52. ① 2 3 4
50. 1 ② 3 4	51. 1 2 ③ 4	54. 1 ② 3 4	54. 1 ② 3 4	54. 1 ② 3 4
54. 1 ② 3 4	54. 1 ② 3 4	55. 1 2 3 ④	55. 1 2 3 ④	55. 1 2 3 ④

FORM L

1. 1 2 3 **4**
4. 1 **2** 3 4
6. **1** 2 3 4
9. 1 **2** 3 4
12. 1 **2** 3 4
15. 1 **2** 3 4
18. 1 2 3 **4**
20. **1** 2 3 4
22. 1 **2** 3 4
24. **1** 2 3 4
26. 1 2 3 **4**
28. **1** 2 3 4
29. 1 2 3 **4**
33. 1 2 **3** 4
34. 1 **2** 3 4
36. 1 **2** 3 4
37. 1 **2** 3 4
38. 1 **2** 3 4
41. **1** 2 3 4
42. **1** 2 3 4
45. 1 2 **3** 4
47. 1 2 **3** 4
49. 1 **2** 3 4
50. 1 **2** 3 4
52. **1** 2 3 4

FORM M

3. 1 2 3 **4**
4. 1 **2** 3 4
8. **1** 2 3 4
11. 1 2 **3** 4
12. 1 **2** 3 4
14. 1 2 **3** 4
19. 1 2 **3** 4
21. 1 2 **3** 4
23. 1 2 3 **4**
26. 1 2 3 **4**
27. 1 2 **3** 4
29. 1 2 3 **4**
31. 1 2 3 **4**
33. 1 2 **3** 4
36. 1 **2** 3 4
39. 1 2 **3** 4
40. 1 2 **3** 4
41. **1** 2 3 4
42. **1** 2 3 4
45. 1 2 **3** 4
50. 1 **2** 3 4
51. 1 2 **3** 4
53. 1 2 **3** 4
54. 1 **2** 3 4
55. 1 2 3 **4**

FORM N

2. 1 **2** 3 4
3. 1 2 3 **4**
9. 1 **2** 3 4
12. 1 **2** 3 4
14. 1 2 **3** 4
15. 1 **2** 3 4
16. **1** 2 3 4
18. 1 2 3 **4**
20. **1** 2 3 4
21. 1 2 **3** 4
24. **1** 2 3 4
28. **1** 2 3 4
29. 1 2 3 **4**
32. **1** 2 3 4
33. 1 2 **3** 4
36. 1 **2** 3 4
37. 1 **2** 3 4
40. 1 2 **3** 4
42. **1** 2 3 4
44. 1 **2** 3 4
47. 1 2 **3** 4
48. 1 **2** 3 4
51. 1 2 **3** 4
53. 1 2 **3** 4
54. 1 **2** 3 4

FORM O

2. 1 **2** 3 4
3. 1 2 3 **4**
5. 1 **2** 3 4
6. **1** 2 3 4
10. 1 2 **3** 4
11. 1 2 **3** 4
13. 1 2 3 **4**
14. 1 2 **3** 4
19. 1 2 **3** 4
21. 1 2 **3** 4
22. 1 **2** 3 4
23. 1 2 3 **4**
28. **1** 2 3 4
30. 1 2 **3** 4
31. 1 2 3 **4**
32. **1** 2 3 4
35. 1 2 3 **4**
36. 1 **2** 3 4
41. **1** 2 3 4
42. **1** 2 3 4
46. 1 2 **3** 4
47. 1 2 **3** 4
50. 1 **2** 3 4
52. **1** 2 3 4
53. 1 2 **3** 4

Appendix F

METRIC CONVERSION TABLES

THESE PREFIXES MAY BE APPLIED
TO ALL SI UNITS

Multiplies and Submultiples	Prefixes	Symbols
1 000 000 000 000 = 10^{12}	tera (tĕr′ å)	T
1 000 000 000 = 10^{9}	giga (jĭ′ gå)	G
1 000 000 = 10^{6}	mega (mĕg′ å)	M *
1000 = 10^{3}	kilo (kĭl′ ō)	k *
100 = 10^{2}	hecto (hĕk′ tō)	h
10 = 10	deka (dĕk′ å)	da
0.1 = 10^{-1}	deci (dĕs′ ĭ)	d
0.01 = 10^{-2}	centi (sĕn′ tĭ)	c *
0.001 = 10^{-3}	milli (mĭl′ ĭ)	m *
0.000 001 = 10^{-6}	micro (mī′ krō)	μ *
0.000 000 001 = 10^{-9}	nano (năn′ ō)	n
0.000 000 000 001 = 10^{-12}	pico (pē′ kō)	p
0.000 000 000 000 001 = 10^{-15}	femto (fĕm′ tō)	f
0.000 000 000 000 000 001 = 10^{-18}	atto (ăt′ tō)	a

*Most commonly used

LINEAR UNITS OF MEASUREMENT

Meter: Originally one ten-millionth (0.000 000 1) of the distance from the equator through Paris to the North Pole. The present standard for the meter is defined in terms of the wavelength of light emitted by the element krypton. The length of the meter is 1 650 763.73 wavelengths of the orange-red line in the spectrum of krypton 86.

VOLUME UNITS OF MEASUREMENT

Liter: Originally the volume described as a cube one decimeter along each edge or 1000 cubic centimeters. The liter was later defined as the volume of 1 kilogram of pure water at a temperature of 4°C. This is approximately 1 000.028 cubic centimeters or nearly one English system quart. In 1964 the 12th General Conference on Weights and Measures redefined the liter to be equal to a cubic decimeter. It is no longer based on the volume of water.

Milliliter: A volume 0.001 of a cubic decimeter.

MASS UNITS OF MEASUREMENT

Gram: The mass comparable to 0.001 of the mass of a platinum-iridium cylinder in Paris, France, however one milliliter of pure water at a temperature of 4°C. is still used for practical purposes.

Kilogram: The mass comparable to a platinum-iridium cylinder stored in Paris, France. The kilogram mass unit is approximately equivalent to a mass having a weight of 2.2 pounds at sea level (however one milliliter of pure water at a temperature of 4°C. is still used for practical purposes).

METRIC UNITS COMPARED:

1 millimeter (mm)	=	0.03937 inches (in.)	1 milligram (mg)	=	0.015 grain
1 centimeter (cm)	=	0.3937 inches	1 centigram (cg)	=	0.154 grain
1 decimeter (dm)	=	3.937 inches	1 gram (g)	=	15.432 grains
1 meter (m)	=	39.37 inches	1 gram	=	1000 milligrams
1 meter	=	1000 millimeters	1 kilogram (kg)	=	1000 grams
1 meter	=	100 centimeters	1 kilogram	=	35.274 ounces (oz)
1 dekameter (dam)	=	393.7 inches	1 kilogram	=	2.204 pounds (lb)
1 hectometer (hm)	=	328 feet (ft)	10 milligrams	=	1 centigram
1 kilometer (km)	=	0.62137 mile (mi)	10 centigrams	=	1 decigram (dg)
1 angstrom (A)	=	0.000 000 000 1 meter (10^{-10} meter)	10 decigrams	=	1 gram
1 micron (μ)	=	0.000 001 meter (10^{-6} meter)	10 grams	=	1 dekagram (dag)

1 liter (l)	=	1.76 pints (pt)	10 dekagrams	=	1 hectogram (hg)
1 liter	=	61.027 cubic inches (in.^{3})	10 hectograms	=	1 kilogram
1 liter	=	1000 milliliters (ml)	1 metric ton (t)	=	1000 kilograms

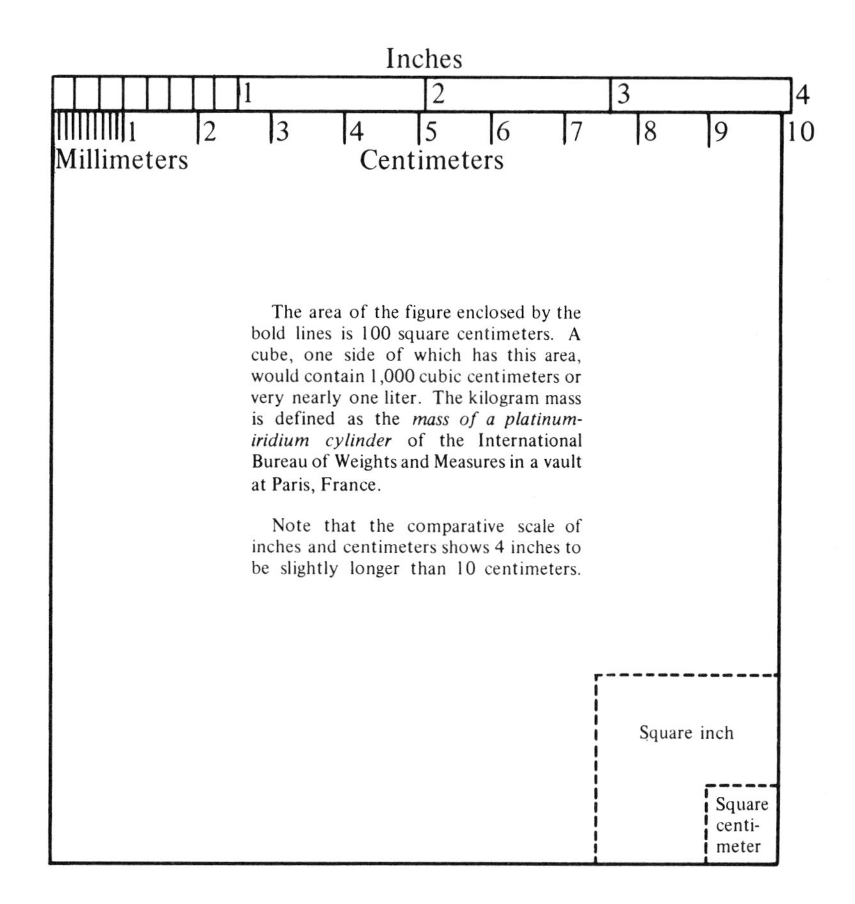

The area of the figure enclosed by the bold lines is 100 square centimeters. A cube, one side of which has this area, would contain 1,000 cubic centimeters or very nearly one liter. The kilogram mass is defined as the *mass of a platinum-iridium cylinder* of the International Bureau of Weights and Measures in a vault at Paris, France.

Note that the comparative scale of inches and centimeters shows 4 inches to be slightly longer than 10 centimeters.

Think Metrically

25 Degrees
Fahrenheit

25 Degrees
Celsius

Appendix G

LABORATORY EQUIPMENT – ILLUSTRATIONS

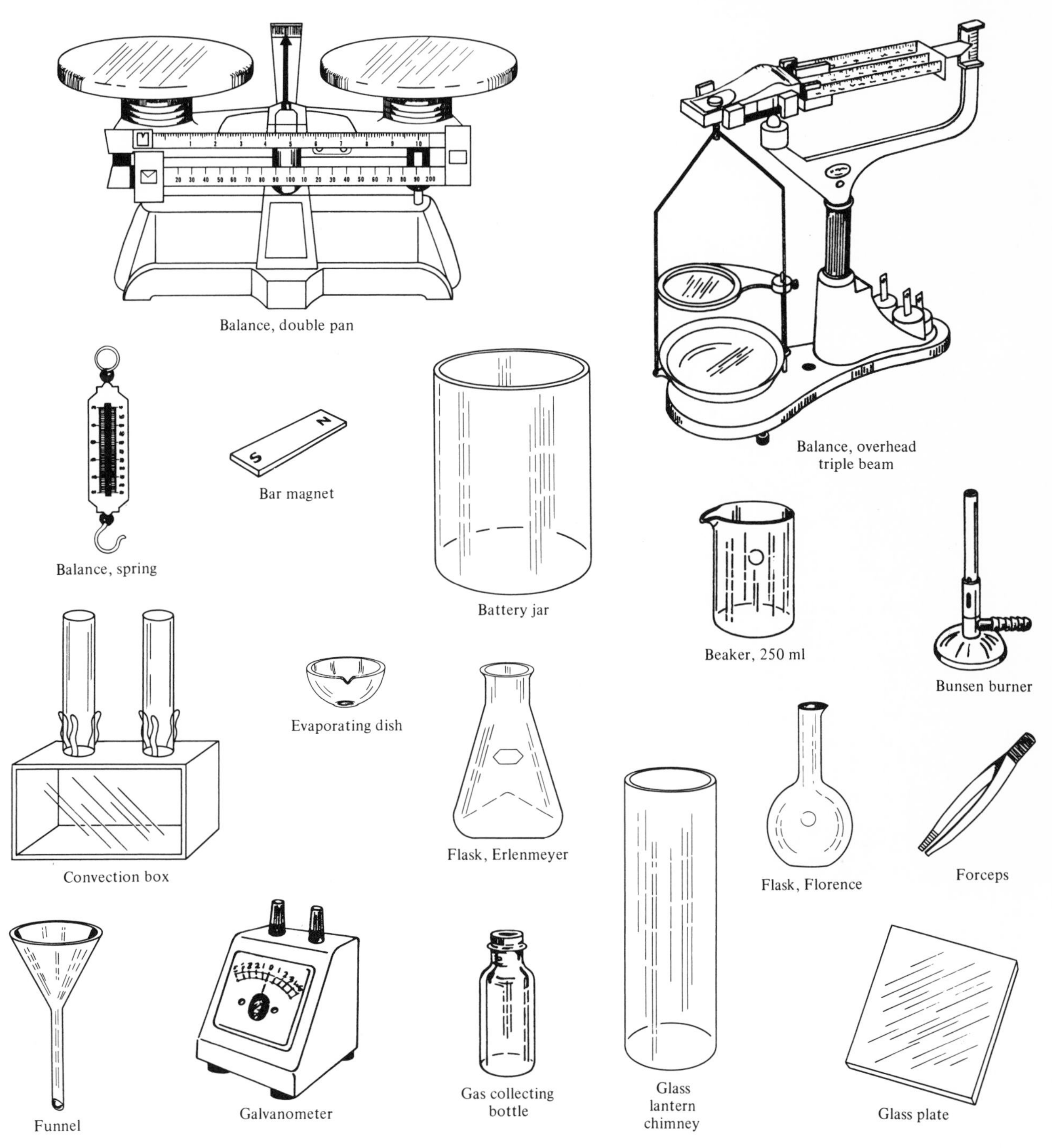

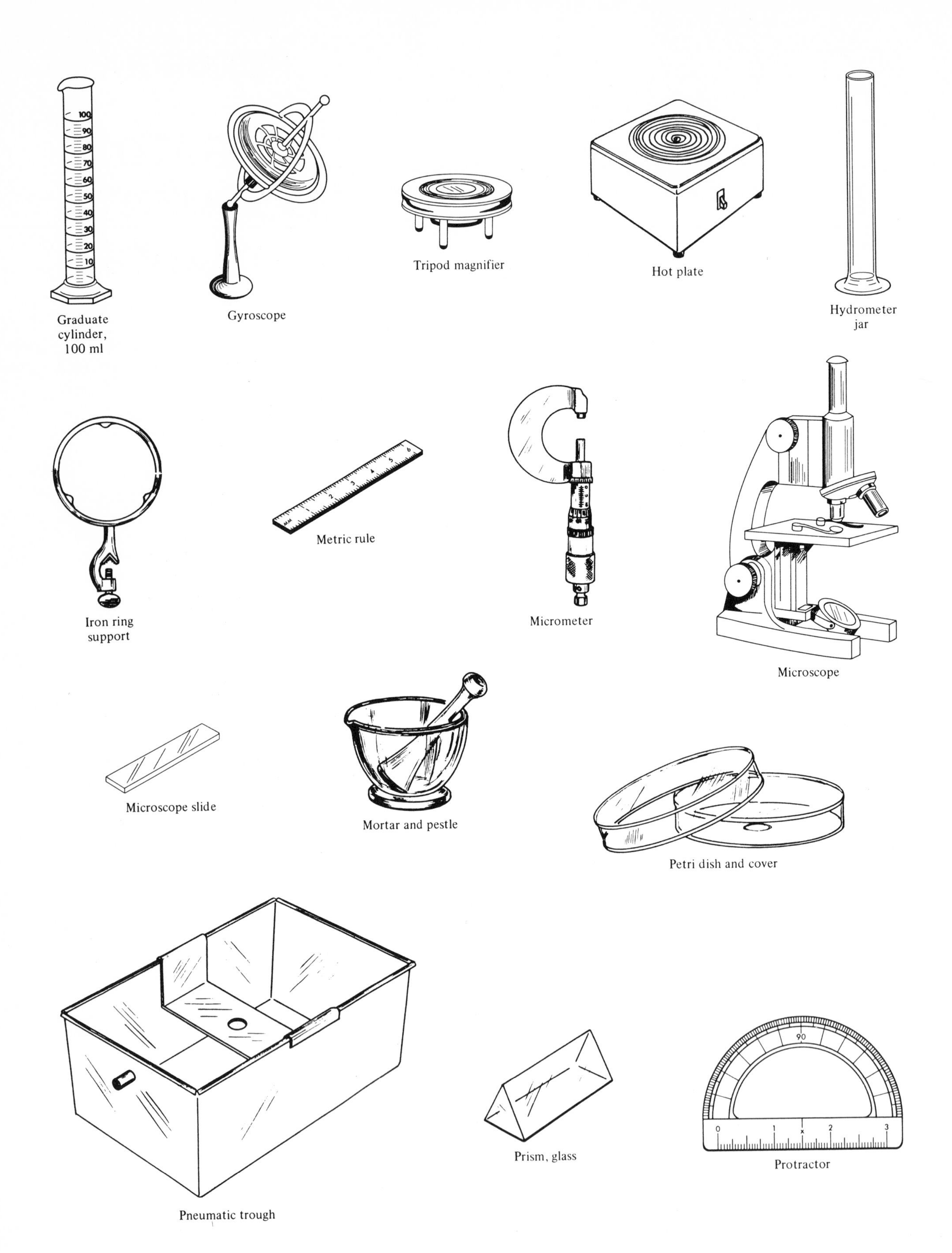
Graduate cylinder, 100 ml
Gyroscope
Tripod magnifier
Hot plate
Hydrometer jar
Iron ring support
Metric rule
Micrometer
Microscope
Microscope slide
Mortar and pestle
Petri dish and cover
Pneumatic trough
Prism, glass
Protractor

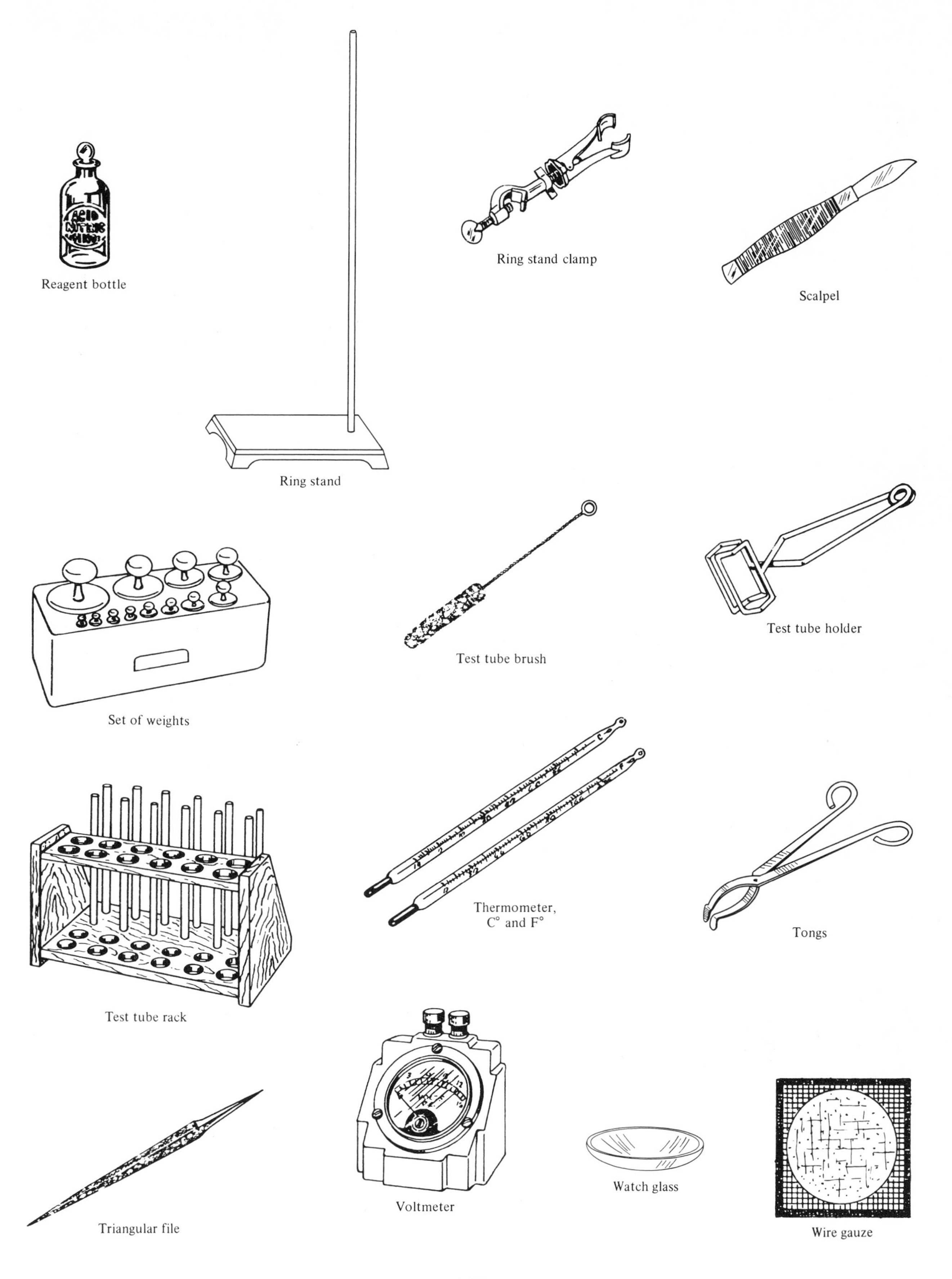
Reagent bottle
Ring stand
Ring stand clamp
Scalpel
Set of weights
Test tube brush
Test tube holder
Test tube rack
Thermometer,
C° and F°
Tongs
Triangular file
Voltmeter
Watch glass
Wire gauze

Appendix H

SAFETY PRACTICES IN THE LABORATORY

These precautions and suggestions are for your own safety and that of your classmates. Study this sheet thoroughly and refer to it when in doubt as to procedure.

1. Report immediately to your teacher any accident, however slight.

2. Observe all signs, labels, and other directions that recommend caution.

3. Ask for further instructions whenever you do not understand clearly what you are to do.

4. Do not handle laboratory equipment, materials, plants, or animals without permission.

5. Take special care to operate any equipment within the limits of pressure, speed, temperature and the like, as recommended.

6. Wrap any glass tubing, bottle, or bulb in wet cloth or paper if any pressure, internal or external, is to be applied to it. When inserting glass rod or tubing in rubber corks, use water to lubricate, then use a gentle, twisting motion when inserting or removing.

7. Dispose of broken glass, chemicals, and other waste only in special receptacles provided for that purpose.

8. Remove glass tubing, thermometer, or other equipment from stoppers immediately after use to avoid *freezing.*

9. Follow carefully instructions for handling any chemical. Never taste any material unless instructed to do so. Use caution when smelling.

10. If an acid or a base is spilled, immediately wash with plenty of water. Notify your teacher. When diluting an acid or a base, add the more concentrated solution to the water, never the water to the concentrated solution.

11. Follow carefully instructions for using any sharp instrument.

12. Avoid breathing directly the fumes of formaldehyde, carbon tetrachloride, ether, chloroform, preservatives, dyes, or other chemicals.

13. Before using any specimen preserved in formaldehyde or other preservative, wash thoroughly in running water. Notify your teacher if any specimens you are using show signs of decay.

14. Know where the first aid kit, fire blanket, fire extinguisher, and related safety equipment are kept in the laboratory.

15. Clean and replace all glassware, apparatus, and instruments after using.